OCR

RECOGNISING ACHIEVEMENT

G C S E
Mathematics

Graduated Assessment

Stages 9 & 10

Authors

Howard Baxter

Mike Handbury

John Jeskins

Jean Matthews

Mark Patmore

Contributor

Colin White

Series editor *Brian Seager*

Hodder & Stoughton

A MEMBER OF THE HODDER HEADLINE GROUP

Orders: please contact Bookpoint Ltd, 130 Milton Park, Abingdon, Oxon OX14 4SB. Telephone: (44) 01235 827720, Fax: (44) 01235 400454. Lines are open from 9.00 – 6.00, Monday to Saturday, with a 24 hour message answering service. Email address: orders@bookpoint.co.uk

British Library Cataloguing in Publication Data

A catalogue record for this title is available from The British Library.

ISBN 0 340 801921

First published 2002

Impression number 10 9 8 7 6 5 4 3

Year 2007 2006 2005 2004 2003 2002

Copyright © 2002 by Howard Baxter, Mike Handbury, John Jeskins, Jean Matthews, Mark Patmore and Brian Seager.

Cover illustration by Mike Stones.

Produced by Hardlines, Charlbury, Oxon.

Printed in Italy for Hodder & Stoughton Educational, a division of Hodder Headline Plc, 338 Euston Road, London NW1 3BH.

Every effort has been made to trace ownership of copyright. The Publishers would be happy to make arrangements with any copyright holder whom it has not been possible to trace.

This book covers the last part of the specification for the Higher tier of GCSE Mathematics. It is particularly aimed at OCR Mathematics C (Graduated Assessment) but could be used for other GCSE Mathematics examinations.

The work in this book covers the criteria in stages M9 and M10, and aims to make the best of your performance in the module tests and the terminal examination:

- Each chapter is presented in a style intended to help you understand the mathematics, with straightforward explanations and worked examples.
- At the start of each chapter is a list of what you should already know before you begin.
- There are plenty of exercises for you to work through and practise the skills.
- At the end of each chapter there is a list of key ideas.
- After every three or four chapters there is a revision exercise.
- Some exercises are designed to be done without a calculator so that you can practise for the non-calculator sections of the papers.
- Many chapters contain Activities to help you develop the necessary skills to undertake coursework.
- At frequent intervals throughout the book there are exam tips, where the experienced examiners who have written this book offer advice and tips to improve your examination performance.
- Revision exercises and Module tests are provided in the Teacher's Resource.

Part of the examination is a calculator-free zone. You will have to do the first section of each paper without a calculator and the questions are designed appropriately.

Most of the marks given for Algebra in AO2 are for 'manipulative' algebra. This includes simplifying algebraic expressions, factorising, solving equations and changing formulae. Some questions are also being set which offer you little help to get started. These are called 'unstructured' or 'multi-step' questions. Instead of the question having several parts, each of which leads to the next, you have to work out the necessary steps to find the answer. There will be examples of this kind of question in the revision tests and past examination papers.

Top ten tips

Here are some general tips from the examiners to help you do well in your tests and examination.

Practise:

1 all aspects of **manipulative algebra** in the specification
2 answering questions **without** a calculator
3 answering questions which require **explanations**
4 answering **unstructured** questions
5 **accurate** drawing and construction
6 answering questions which **need a calculator**, trying to use it efficiently
7 **checking answers**, especially for reasonable size and degree of accuracy
8 making your work **concise** and well laid out
9 using the **formula sheet** before the examination
10 **rounding** numbers, but only at the appropriate stage.

Coursework

The GCSE Mathematics examinations will assess your ability to use your mathematics on longer problems than those normally found on timed written examination papers. Assessment of this type of work will account for 20% of your final mark. It will involve two tasks, each taking about three hours. One task will be an investigation, the other a statistics task.

Each type of task has its own mark scheme in which marks are awarded in three categories or 'strands'. The titles of these strands give you clues about the important aspects of this work.

For the investigation tasks the strands are:

● Making and monitoring decisions – what you are going to do and how you will do it
● Communicating mathematically – explaining and showing exactly what you have done
● Developing the skills of mathematical reasoning – using mathematics to analyse and prove your results.

The table below gives some idea of what you will have to do and show. Look at this table whenever you are doing some extended work and try to include what it suggests you do.

Mark	Making and monitoring decisions	Communicating mathematically	Developing the skills of mathematical reasoning
1	organising work, producing information and checking results	discussing work using symbols and diagrams	finding examples that match a general statement
2	beginning to plan work, choosing your methods	giving reasons for choice of presentation of results and information	searching for a pattern using at least three results
3	finding out necessary information and checking it	showing understanding of the task by using words, symbols, diagrams	explaining reasoning and making a statement about the results found

Mark	Making and monitoring decisions	Communicating mathematically	Developing the skills of mathematical reasoning
4	simplifying the task by breaking it down into smaller stages	explaining what the words, symbols and diagrams show	testing generalisations by checking further cases
5	introducing new questions leading to a fuller solution	justifying the means of presentation	justifying solutions explaining why the results occur
6	using a range of techniques and reflecting on lines of enquiry and methods used	using symbolisation consistently	explaining generalisations and making further progress with the task
7	analysing lines of approach and giving detailed reasons for choices	using symbols and language to produce a convincing and reasoned argument	report includes mathematical justifications and explanations of the solutions to the problem
8	exploring extensively an unfamiliar context or area of mathematics and applying a range of appropriate mathematical techniques to solve a complex task	using mathematical language and symbols efficiently in presenting a concise reasoned argument	providing a mathematically rigorous justification or proof of the solution considering the conditions under which it remains valid

For the statistical tasks the strands are:

- Specifying the problem and planning – choosing or defining a problem and outlining the approach to be followed
- Collecting, processing and representing data – explaining and showing what you have done
- Interpreting and discussing results – using mathematical and statistical knowledge and techniques to analyse, evaluate and interpret your results and findings.

The marks obtained from each task are added together to give a total out of 48.

The table below gives some idea of what you will have to do and show. Look at this table whenever you are doing some extended work and try to include what it suggests you do.

Mark	Specifying the problem and planning	Collecting, processing and representing data	Interpreting and discussing results
1–2	choosing a simple problem and outlining a plan	collecting some data; presenting information, calculations and results	making comments on the data and results
3–4	choosing a problem which allows you to use simple statistics and plan the collection of data	collecting data and then processing it using appropriate calculations involving appropriate techniques; explaining what the words, symbols and diagrams show	explaining and interpreting the graphs and calculations and any patterns in the data

Mark	Specifying the problem and planning	Collecting, processing and representing data	Interpreting and discussing results
5–6	considering a more complex problem and using a range of techniques and reflecting on the method used	collecting data in a form that ensures they can be used; explaining statistical meaning through the consistent use of accurate statistics and giving a reason for the choice of presentation; explaining features selected	commenting on, justifying and explaining results and calculations; commenting on the methods used
7–8	analysing the approach and giving reasons for the methods used; using a range of appropriate statistical techniques to solve the problem	using language and statistical concepts effectively in presenting a convincing reasoned argument; using an appropriate range of diagrams to summarise the data and show how variables are related	correctly summarising and interpreting graphs and calculations and making correct and detailed inferences from the data; appreciating the significance of results obtained and, where relevant, allowing for the nature and size of the sample and any possible bias; evaluating the effectiveness of the overall strategy and recognising limitations of the work done, making suggestions for improvement

Advice

Starting a task

Ask yourself:
- what does the task tell me?
- what does it ask me?
- what can I do to get started?
- what equipment and materials do I need?

Working on the task

- Make sure you explain your method and present your results as clearly as possible
- Break the task down into stages. For example in 'How many squares on a chessboard', begin by looking at 1 × 1 squares then 2 × 2 squares, then 3 × 3 squares. In a task asking for the design of a container, start with cuboids then nets, surface area, prisms … Or in statistics you might want to start with a pilot survey or questionnaire.
- Write down questions that occur to you, for example, *what happens if you change the size of a rectangle systematically?* They may help you find out more about the work. In a statistical task you might wish to include different age groups or widen the type of data.

- Explore as many aspects of the task as possible.
- Develop the task into new situations and explore these thoroughly.
    - What connections are possible?
    - Is there a result to help me?
    - Is there a pattern?
    - Can the problem be changed? If so, how?

Explain your work

- Use appropriate words and suitable tables, diagrams, graphs, calculations.
- Link as much of your work together as possible, explaining, for example, why you chose the tables and charts you used and rejected others, or why the median is more appropriate than the mean in a particular statistical analysis, or why a pie chart is not appropriate. Don't just include diagrams to show identical information in different ways.
- Use algebra or symbols to give clear and efficient explanations; in investigations, you must use algebra to progress beyond about 4 marks. You will get more credit for writing $T = 5N + 1$ than for writing 'the total is five times the pattern number, plus one'.
- Don't waffle or use irrelevant mathematics; present results and conclusions clearly.

State your findings

- Show how patterns have been used and test conclusions.
- State general results in words and explain what they mean.
- Write formulae and explain how they have been found from the situations explored.
- Prove the results using efficient mathematical methods.
- Develop new results from previous work and use clear reasoning to prove conclusions.
- Make sure your reasoning is accurate and draws upon the evidence you've presented.
- Show findings in clear, relevant diagrams.
- Check you've answered the question or hypothesis.

Review/conclusion/extension

- Is the solution acceptable?
- Can the task be extended?
- What can be learned from it?

Example task

On the next page there is a short investigative task for you to try, in both 'structured' and 'unstructured' form. The structured form shows the style of a question that might appear on a timed written paper. The unstructured form represents the usual style of a coursework task. The structured form leads you to an algebraic conclusion. Notice the appearance of algebra from question 4 onwards, through a series of structured questions. These mirror the sort of questions you would be expected to think of (and answer) if you were trying it as coursework.

Comments about the questions, linking the two forms of presentation, are also shown.

Although the task in both forms directs you to investigate trapezium numbers, you would be expected to extend the investigation into other forms of number, such as pentagon numbers, to achieve the higher marks.

ACTIVITY

structured form

Trapezium numbers

These diagrams represent the first three trapezium numbers.

Each diagram always starts with two dots on the top row.

1st	2nd	3rd
••	•• •••	•• ••• ••••
2 dots	5 dots	9 dots

So the third trapezium number is 9 because nine dots can be arranged as a trapezium. There are two dots in the top row, three dots in the next row and four dots in the bottom row.

1 Write down the next two trapezium numbers

2 a) Draw a table, graph or chart of all the trapezium numbers, from the first to the tenth.
 b) Work out the eleventh trapezium number.

3 The 19th trapezium number is 209. Explain how you could work out the 20th trapezium number without drawing any diagrams.

4 Find an expression for the number of dots in the bottom row of the nth trapezium number.
Test your expression for a suitable value of n.

5 Find, giving an explanation, an expression for the number of dots in the bottom row of the diagram for the $(n + 1)$th trapezium number.

6 The nth trapezium number is x. Write down an expression in terms of x and n for the $(n + 1)$th trapezium number. Test your expression for a suitable value of n.

unstructured form

Trapezium numbers

These diagrams represent the first three trapezium numbers.

Each diagram starts with two dots on the top row.

1st	2nd	3rd
••	•• •••	•• ••• ••••
2 dots	5 dots	9 dots

So the third trapezium number is 9 because nine dots can be arranged as a trapezium.

Investigate trapezium numbers

NB Although the task in this form asks you to investigate trapezium numbers, you have the freedom to – and are expected to – extend the investigation to consider other forms of number such as pentagon numbers.

Commentary

This question allows you to show understanding of the task, systematically obtaining information which **could** enable you to find an expression for trapezium numbers.

This question provides a structure, using symbols, words and diagrams, from which you should be able to derive an expression from either a table or a graph. Part **b)** could be done as a 'predict and test'.

In the unstructured form you would not normally answer a question like this.

From here you are **directed** in the structured task, and **expected** in the unstructured task, to use algebra, testing the expression – the **generalisation**.

In the unstructured form this would represent the sort of 'new' question you might ask, to lead to a further solution and to demonstrate symbolic presentation and the ability to relate the work to the physical structure, rather than doing all the analysis from a table of values.

Stage 9

CONTENTS

1 Checking answers

You should already know

- how to interpret significant figures
- how to write numbers in standard form.

Checking answers by rounding to one significant figure

It is important to be able to check calculations quickly, without using a calculator. One way to do this is to round the numbers to one significant figure.

EXAMPLE 1

Find an approximate answer to the calculation $5 \cdot 13 \times 4 \cdot 83$.

$5 \cdot 13 \times 4 \cdot 83 = 5 \times 5$

Rounding $5 \cdot 13$ and $4 \cdot 83$ each to one significant figure to give a much simpler calculation

$= 25.$

Exam tip

In a calculation it may be possible to round one number up and another number down. This might give an answer close to the exact answer.

EXAMPLE 2

Find an approximate answer to the calculation
$$\frac{(3 \cdot 26 \times 10^3) \times (8 \cdot 17 \times 10^5)}{(6 \cdot 28 \times 10^2)}$$

$$\approx \frac{3 \times 8}{6} \times \frac{10^3 \times 10^5}{10^2}$$

Round numbers to 1 significant figure.
Collect together numbers and powers of 10.

$$\approx \frac{24}{6} \times \frac{10^8}{10^2}$$

Add indices when multiplying powers of 10.

$$\approx 4 \times 10^6$$

Subtract indices when dividing powers of 10.

EXERCISE 1.1A

1 Find approximate answers to these calculations by rounding each number to one significant figure.

a) $498 \times 2 \cdot 18$ **b)** $13 \cdot 92 \div 6 \cdot 8$

c) $4 \cdot 19 \times 6 \cdot 68$ **d)** $881 \div 99$

Now use a calculator to see how close your approximations are to the correct answers.

2 Find approximate answers to these calculations by rounding each number to one significant figure.

a) $159 \cdot 65 \div 515$ **b)** $36 \cdot 8 \times (5 \cdot 7 + 6 \cdot 4)$ **c)** $\sqrt{41300}$

Now use a calculator to see how close your approximations are to the correct answers.

3 Find approximate answers to these calculations by rounding each number to one significant figure.

a) $\dfrac{2 \cdot 5 \times 3 \cdot 6}{5 \cdot 9}$ **b)** $\dfrac{0 \cdot 21 \times 93}{103 \cdot 1 \div 9 \cdot 6}$

c) $3 \cdot 8 \times \sqrt{385}$ **d)** $\dfrac{543}{18 \cdot 1} + \dfrac{472}{10 \cdot 9}$

4 Find approximate answers to these calculations by rounding each number to one significant figure.

a) $(1 \cdot 98 \times 10^5) \times (4 \cdot 65 \times 10^4)$ **b)** $(1 \cdot 5 \times 10^8) \times (7 \cdot 2 \times 10^{-4})$

c) $\dfrac{(7 \cdot 89 \times 10^5)}{(4 \cdot 73 \times 10^3)}$ **d)** $(5 \cdot 59 \times 10^2) \div (1 \cdot 87 \times 10^5)$

e) $\dfrac{(5 \cdot 84 \times 10^4)}{(2 \cdot 68 \times 10^{-2})}$ **f)** $\dfrac{(8 \cdot 27 \times 10^{13}) \times (9 \cdot 75 \times 10^2)}{(1 \cdot 25 \times 10^8)}$

g) $\dfrac{(6 \cdot 89 \times 10^5) \times (7 \cdot 36 \times 10^{-4})}{(4 \cdot 57 \times 10^{-3})}$ **h)** $\dfrac{(1 \cdot 25 \times 10^5)^2}{(3 \cdot 6 \times 10^4)}$

Now use a calculator to see how close your approximations are to the correct answers.

EXERCISE 1.1B

1 Find approximate answers to these calculations by rounding each number to one significant figure.

a) $7·2 \times 9·7$ **b)** $105·6 \div 5·12$ **c)** $313 \times 0·68$ **d)** $4·189 \div 0·477$

Now use a calculator to see how close your approximations are to the correct answers.

2 Find approximate answers to these calculations by rounding each number to one significant figure.

a) $0·143 \div 0·116$ **b)** $(5·67 - 3·85) \times 39$ **c)** $(34·2)^2$

Now use a calculator to see how close your approximations are to the correct answers.

3 Find approximate answers to these calculations by rounding each number to one significant figure.

a) $\dfrac{28·2 \times 3·14}{8·99}$ **b)** $96·7 \times 4·9^2$ **c)** $\dfrac{54·3 + 47·2}{9·8 + 10·9}$ **d)** $\dfrac{\sqrt{5·21 \times 8·35 \times 0·105}}{1·72^2}$

4 Find approximate answers to these calculations by rounding each number to one significant figure.

a) $(3·2 \times 10^6) \times (9·45 \times 10^4)$ **b)** $(3·64 \times 10^7) \times (2·4 \times 10^{-5})$

c) $\dfrac{(5·93 \times 10^5)}{(3·29 \times 10^4)}$ **d)** $(3·52 \times 10^4) \div (1·44 \times 10^8)$

e) $\dfrac{(8·17 \times 10^{-3})}{(1·52 \times 10^{-2})}$ **f)** $\dfrac{(4·29 \times 10^{-3}) \times (8·18 \times 10^{-5})}{(1·5 \times 10^3)}$

g) $\dfrac{(3·75 \times 10^9)}{(5·01 \times 10^{-3}) \times (1·62 \times 10^6)}$ **h)** $\dfrac{(4·17 \times 10^{-3}) \times (9·29 \times 10^{-5})}{\sqrt{(8·63 \times 10^8)}}$

Now use a calculator to see how close your approximations are to the correct answers.

Key idea

● Check answers to calculations by rounding numbers to one significant figure.

2 Algebraic manipulation

You should already know

- how to factorise simple expressions
- how to expand brackets and manipulate simple algebraic expressions
- how to factorise simple quadratic equations such as $x^2 + bx + c$.

Multiplying out two brackets

Expressions such as $a(3a - 2b)$ can be multiplied out to give
$a(3a - 2b) = (a \times 3a) - (a \times 2b) = 3a^2 - 2ab$.

This can be extended to working out expressions such as
$(2a + b)(3a + b)$.

Each term of the first bracket must be multiplied by each term
of the second bracket.

$(2a + b)(3a + b)$

$= 2a(3a + b) + b(3a + b)$ Expanding the first bracket.

$= 6a^2 + 2ab + 3ab + b^2$ Notice that the middle two terms are **like terms** and so can be collected.

$= 6a^2 + 5ab + b^2$

Exam tip

Most errors are made in multiplying out the second bracket when the sign in front is negative.

Exam tip

Apart from multiplying out brackets, you may sometimes be asked to simplify, expand or remove the brackets, which all mean the same thing.

EXAMPLE 1

Multiply out the brackets.

a) $(2a + 3)(a - 1)$ **b)** $(5a - 2b)(3a - b)$ **c)** $(2a - b)(a + 2b)$

a) $(2a + 3)(a - 1) = 2a(a - 1) + 3(a - 1) = 2a^2 - 2a + 3a - 3$

$\quad\quad\quad\quad = 2a^2 + a - 3$

> Be careful with the signs.

b) $(5a - 2b)(3a - 2b) = 5a(3a - b) - 2b(3a - b)$

$\quad\quad\quad\quad\quad = 15a^2 - 5ab - 6ab + 2b^2$

$\quad\quad\quad\quad\quad = 15a^2 - 11ab + 2b^2$

> Note that it is ^-2b times the bracket.

c) $(2a - b)(a + 2b) = 2a(a + b) - b(a + 2b)$

$\quad\quad\quad\quad\quad = 2a^2 + 4ab - ab - 2b^2$

$\quad\quad\quad\quad\quad = 2a^2 + 3ab - 2b^2$

> Note that it is ^-b times the bracket.

Exam tip

Take care with negative signs.

In each part of example 1, the two brackets have resulted in three terms.

There are two other types of expansions of two brackets that you need to know about.

EXAMPLE 2

Exam tip

The important thing in Example 2a is to make sure that you write the brackets separately and that you end up with three terms.

Expand the brackets.

a) $(2a - 3b)^2$ **b)** $(2a - b)(2a + b)$

a) $(2a - 3b)^2 = (2a - 3b)(2a - 3b)$

$\quad\quad\quad\quad = 2a(2a - 3b) - 3b(2a - 3b) = 4a^2 - 6ab - 6ab + 9b^2$

$\quad\quad\quad\quad = 4a^2 - 12ab + 9b^2$

> Note that $^-6ab - 6ab = {}^-12ab$

b) $(2a - b)(2a + b) = 2a(2a + b) - b(2a + b) = 4a^2 + 2ab - 2ab - b^2$

$\quad\quad\quad\quad\quad = 4a^2 - b^2$

> Note that we only get two terms here because the middle terms cancel each other out.
> This type is known as the difference of two squares because: $(a - b)(a + b) = a^2 - b^2$

Exam tip

Some people can multiply out two brackets without writing down anything. However, you are more likely to make an error by missing steps and so it is worth showing every step in an examination.

Chapter 2 Algebraic manipulation

EXERCISE 2.1A

Multiply out the brackets.

1 $(x + 2)(x - 3)$ **6** $(5x - 1)(2x - 4)$
2 $(x + 5)(x + 9)$ **7** $(4x + 2)(3x - 7)$
3 $(x - 1)^2$ **8** $(x + y)(2x + y)$
4 $(5 + x)(x - 6)$ **9** $(3x - 5y)(x - 4y)$
5 $(x - 7)(x + 7)$ **10** $(3x + 4y)(4x - 5y)$

EXERCISE 2.1B

Multiply out the brackets.

1 $(x + 2)(x + 1)$ **6** $(2x - 5)(3x - 2)$
2 $(x - 3)(x + 6)$ **7** $(5x + 6)(2x - 3)$
3 $(x - 4)^2$ **8** $(3x + y)(4x + y)$
4 $(2 + x)(5 + x)$ **9** $(2x - 3y)(x - 2y)$
5 $(x - 8)(x + 8)$ **10** $(7x + 8y)(6x - 4y)$

Simplifying expressions using indices

Remember that: $\quad a \times a \times a = a^3$ and $a \times a \times a \times a \times a = a^5$.

This can be extended: $a^3 \times a^5 = (a \times a \times a) \times (a \times a \times a \times a \times a) = a^8$

which is the same as: $a^3 \times a^5 = a^{3 + 5} = a^8$.

This suggests a general rule for indices.

$a^m \times a^n = a^{m + n}$

Similarly: $a^5 \div a^3 = (a \times a \times a \times a \times a) \div (a \times a \times a)$

$$= \frac{a \times a \times a \times a \times a}{a \times a \times a} = a \times a = a^2$$

Cancelling $a \times a \times a$ top and bottom.

This is the same as $a^5 \div a^3 = a^{5 - 3} = a^2$

which suggests another general rule for indices.

$a^m \div a^n = a^{m - n}$

Now $(a^2)^3 = a^2 \times a^2 \times a^2 = a^6$ By the first rule.

This is the same as $(a^2)^3 = a^{2 \times 3} = a^6$

and this suggests yet another rule.

$(a^n)^m = a^{n \times m}$

$a^3 \div a^3 = a^{3-3} = a^0$ but $a^3 \div a^3 = 1$

This gives another rule.

$a^0 = 1$

You can use these rules, together with the algebra you have already learnt, to simplify a number of different algebraic expressions.

EXAMPLE 3

Simplify these.

a) $3a^2 \times 4a^3$

b) $\dfrac{6a^5}{2a^3}$

c) $(a^3)^4 \times a^3 \div a^5$

a) $3a^2 \times 4a^3 = 12a^5$

> The numbers are just multiplied and the indices are added.

b) $\dfrac{6a^5}{2a^3} = 3a^2$

> The numbers are divided and the indices are subtracted.

c) $(a^3)^4 \times a^3 \div a^5 = a^{12} \times a^3 \div a^5$
$= a^{15} \div a^5$
$= a^{10}$

> Use the three rules.

EXAMPLE 4

Simplify where possible.

a) $4a^2b^3 \times 3ab^2$ **b)** $\dfrac{12ab^3 \times 3a^2b}{2a^3b^2}$

c) $4a^2 + 3a^3$

a) $4a^2b^3 \times 3ab^2 = 12a^3b^5$

> The numbers are multiplied and the indices are added for each letter. Note that a is the same as a^1, so $a^2 \times a = a^{2+1} = a^3$.

b) $\dfrac{12ab^3 \times 3a^2b}{2a^3b^2} = 18b^2$

> The numbers combine as $12 \times 3 \div 2 = 18$, $a \times a^2 \div a^3 = a^{1+2-3} = a^0 = 1$, $b^3 \times b \div b^2 = b^{3+1-2} = b^2$.

c) $4a^2 + 3a^3$

> This cannot be simplified. The two terms are different and cannot be added.

7

EXERCISE 2.2A

Simplify where possible.

1 $3a^2 \times 4a^3$

2 $\dfrac{12a^5}{6a^3}$

3 $(3a^3)^2$

4 $2a^2b \times 3a^3b^2$

5 $4a^2b - 2ab^2$

6 $\dfrac{15a^2b^3 \times 3a^2b}{9a^3b^2}$

7 $\dfrac{9p^2q \times (2p^3q)^2}{12p^5q^3}$

8 $\dfrac{4abc \times 3a^2bc^3}{6a^2bc^2}$

9 $\dfrac{12t^3}{(2t)^2}$

10 $2a^2b \times 3ab^2 - 4a^3b^3$

EXERCISE 2.2B

Simplify where possible.

1 $\dfrac{a^5 \times a^3}{a^6}$

2 $3a^2 \times 4a^2$

3 $(2c)^3$

4 $3a^2b^3 \times 2a^3b^4$

5 $12a^2 \times 3b^2$

6 $2a^2 + 3a^3$

7 $8a^2b^3 \times 2a^3b \div 4a^4b^2$

8 $\dfrac{(3a^2b^2)^3}{(a^3b)^2}$

9 $4a^2 \times 2b^3 - a \times 3b \times ab^2$

10 $6a^2 \times (2ab^2)^2 \div 12b^2a$

Factorising algebraic expressions

Factors are numbers or letters which will divide into an expression.

The factors of 6 are 1, 2, 3 and 6.

The factors of b^3 are 1, b, b^2 and b^3.

Remember that multiplying or dividing by 1 leaves a number unchanged, so 1 is not a useful factor and is ignored.

To factorise an expression, look for common factors, e.g. the common factors of $2a^2$ and $6a$ are 2, a and $2a$.

EXAMPLE 5

Factorise these fully.

a) $4p + 6$ **b)** $2a^2 - 3a$

c) $15ab^2 + 10a^2b^2$ **d)** $2a - 10a^2 + 6a^3$

a) $4p + 6 = 2(2p + 3)$

> The only common factor is 2 and
> $2 \times 2p = 4p$, $2 \times 3 = 6$.

b) $2a^2 - 3a = a(2a - 3)$

> The only common factor is a and
> $a \times 2a = 2a^2$, $a \times {}^-3 = {}^-3a$.

c) $15ab^2 + 10a^2b^2 = 5ab^2(3 + 2a)$

> 5, a and b^2 are common factors and
> $5ab^2 \times 3 = 15ab^2$, $5ab^2 \times 2a = 10a^2b^2$.

d) $2a - 10a^2 + 6a^3 = 2a(1 - 5a + 3a^2)$

> 2 and a are common factors and
> $2a \times 1 = 2a$, $2a \times {}^-5a = {}^-10a^2$ and
> $2a \times 3a^2 = 6a^3$.

Exam tip

Make sure that you have found all the common factors. Check that the expression in the bracket will not factorise further.

EXERCISE 2.3A

Factorise these fully.

1 $2a + 8$

2 $3a + 5a^2$

3 $2ab - 6ac$

4 $5a^2b + 10ab^2$

5 $2x^2y^2 - 3x^3y$

6 $3a^2b - 6ab^2$

7 $12x - 6y + 8z$

8 $9ab + 6b^2$

9 $4a^2c - 2ac^2$

10 $15xy - 5y$

11 $6a^3 - 4a^2 + 2a$

12 $3a^2b - 9a^3b^2$

13 $5a^2b^2c^2 - 10abc$

14 $2a^2b - 3a^2b^3 + 7a^4b$

15 $4abc - 3ac^2 + 2a^2b$

Chapter 2   *Algebraic manipulation*

EXERCISE 2.3B

Factorise these fully.

1 $3x - 12$

2 $4a + 5ab$

3 $4ab - 2a^2$

4 $3ab - 2ac + 3ad$

5 $5x^2 - 15x + 15$

6 $4a^2b - 3ab^2$

7 $9x^2y - 6xy^2$

8 $14a^2 - 8a^3$

9 $21x^2 - 14y^2$

10 $12x^2y + 8xy - 4xy^2$

11 $14s^2t - 7st^2$

12 $10z^3 - 15z^2 + 5z$

13 $5abc - 15a^2b^2c^2$

14 $3a^2bc - 6ab^2c - 9abc^2$

15 $7a^3b^3c^2 - 14a^2b^3c^3$

Difference of two squares

You may remember from earlier work that expressions of the type

$x^2 - b^2$

can be factorised into the two brackets

$(x - b)(x + b)$.

(Check that these are the same by multiplying out the brackets and simplifying your answer.)

Exam tip

This method of factorising is important when there is no 'x-term' in a quadratic expression.

EXAMPLE 6

Factorise $x^2 - 16$.

$x^2 - 16 = x^2 - 4^2$

$\qquad = (x - 4)(x + 4)$

In fact *any* expression which can be written as two squares subtracted can be factorised in this way.

EXAMPLE 7

Factorise

a) $25x^2 - 1$ **b)** $9x^2 - 4y^2$

a) $25x^2 - 1 = 5^2x^2 - 1^2$

$\qquad\qquad = (5x - 1)(5x + 1)$

b) $9x^2 - 4y^2 = 3^2x^2 - 2^2y^2$

$\qquad\qquad = (3x - 2y)(3x + 2y)$

When the 'number term' is not a square number, check to see if a common factor needs to be extracted to get the correct form to use the 'difference of two squares' method.

EXAMPLE 8

Factorise $3x^2 - 12$

$$3x^2 - 12 = 3(x^2 - 4)$$
$$= 3(x^2 - 2^2)$$
$$= 3(x - 2)(x + 2)$$

EXERCISE 2.4A

Factorise the following.

1 $x^2 - 25$

2 $4a^2 - b^2$

3 $25x^2 - 49y^2$

4 $a^2 - 9b^2$

5 $100x^2 - 1$

6 $x^2y^2 - 16a^2$

7 $121x^2 - 144y^2$

8 $8 - 2x^2$

9 $7a^2 - 63b^2$

10 $25x^2y^2 - 100$

EXERCISE 2.4B

Factorise the following.

1 $x^2 - 4$

2 $9 - 16y^2$

3 $9x^2 - 64$

4 $1 - 49t^2$

5 $25 - 4x^2$

6 $y^2 - 169$

7 $81p^2 - 36q^2$

8 $3x^2 - 192$

9 $45 - 20x^2$

10 $x^2y^2z^2 - 100$

Factorising quadratic expressions where the coefficient of $x^2 \neq 1$

You have already learnt how to factorise simple quadratic factors such as $x^2 + bx + c$.

Remember:

- if c is positive find two numbers that multiply to c and add up to b
- if c is negative find two numbers that multiply to c and whose difference is b.

The expression $ax^2 + bx + c$, when factorised, will be
$(px + q)(rx + s) = prx^2 + (ps + qr)x + qs$
when expanded.

So $pr = a$, $qs = c$ and $ps + qr = b$. As before, the sign of c defines whether you are looking for a sum or a difference.

It is easiest to look at examples.

EXAMPLE 9

Factorise $3x^2 + 11x + 6$.

As the last sign is +, both signs in the bracket are the same, and as the middle sign is +, they are both +.

The only numbers that can multiply to give 3 are 3 and 1.

So, as a start, $(3x +)(x +)$.

The numbers that multiply to give 6 are either 3 and 2 or 6 and 1.

So the possible answers are

$(3x + 2)(x + 3)$ or $(3x + 3)(x + 2)$ or $(3x + 6)(x + 1)$ or $(3x + 1)$ or $(3x + 1)(x + 6)$.

By expanding the brackets it can be seen that the first one is correct.

The coefficient of the middle term is $3 \times 3 + 2 \times 1 = 9 + 2 = 11$.

It is very useful to check completely by multiplying out the whole bracket.

Writing out all the possible brackets can be a long process and it is quicker to test the possibilities for the middle term until you find the correct one and then multiply out the brackets to check.

EXAMPLE 10

Factorise $4x^2 - 14x + 6$.

The first thing to check is if there is any common factor.

$2(2x^2 - 7x + 3)$

Here the common factor is 2.

Now look at the quadratic expression.

Here you can see that both signs are − and that everything is added.

The possibilities for the first terms in the brackets are 2 and 1 and for the second terms they are 3 and 1 or 1 and 3. The middle term is thus $2 \times 1 + 1 \times 3 = 5$ or $2 \times 3 + 1 \times 1 = 7$. The second is correct so the quadratic factorises to $(2x - 1)(x - 3)$ and the full answer is $2(2x - 1)(x - 3)$.

EXAMPLE 11

Factorise $5x^2 + 13x + 6$.

There is no common factor.

Both signs in the brackets are +.

First terms are 5 and 1, second are 1 and 6, 6 and 1, 3 and 2, or 2 and 3.

If you try $5 \times 1 + 1 \times 6 = 11$, $5 \times 6 + 1 \times 1 = 31$, $5 \times 3 + 1 \times 2 = 17$, $5 \times 2 + 1 \times 3 = 13$.

So the answer is $(5x + 3)(x + 2)$. Check by multiplying out.

EXAMPLE 12

Factorise $6x^2 - 17x + 12$.

There is no common factor.

Both signs in the brackets are −.

First terms are 6 and 1 or 3 and 2, second terms are 1 and 12 or 12 and 1 or 2 and 6 or 6 and 2 or 4 and 3 or 3 and 4. This could mean 12 possible products but if you look at the middle term and see it is 17 you can gather that you will not be multiplying anything by 12 and are unlikely to multiply anything by 6. Try the most likely ones first.

$3 \times 4 + 2 \times 3 = 18$, $3 \times 3 + 2 \times 4 = 17$ which is correct.

So the answer is $(3x - 4)(2x - 3)$. Check by multiplying out.

Exam tip

First look for any common factor then try the most obvious pairs first and remember, if the sign of c is positive, everything is added and both brackets have the same sign as b.

Chapter 2 *Algebraic manipulation*

EXERCISE 2.5A

Factorise the following.

1 $x^2 + 7x + 6$

2 $x^2 - 6x + 8$

3 $2x^2 + 6x + 4$

4 $2x^2 + 9x + 4$

5 $6x^2 - 15x + 6$

6 $3x^2 - 11x + 6$

7 $3x^2 - 11x + 10$

8 $4x^2 + 8x + 3$

9 $5x^2 - 13y + 6$

10 $6x^2 - 19x + 10$

EXERCISE 2.5B

Factorise the following.

1 $x^2 + 5x + 6$

2 $x^2 - 7x + 10$

3 $3x^2 + 7x + 2$

4 $2x^2 + 7x + 6$

5 $3x^2 - 12x + 12$

6 $3x^2 - 13x + 10$

7 $4x^2 - 16x + 15$

8 $7x^2 + 10x + 3$

9 $5x^2 - 22x + 8$

10 $8x^2 - 18x + 9$

The examples looked at so far had c positive and so were more straightforward. The ones where c is negative have different signs in the brackets and the middle term is the difference of the products. These are, again, best shown by examples.

EXAMPLE 13

Factorise $3x^2 - 7x - 6$.

There is no common factor.

The signs are different.

First terms are 3 and 1, second terms are 1 and 6, 6 and 1, 3 and 2, or 2 and 3.

Middle term products are $3 \times 1 - 1 \times 6 = 3$, $3 \times 6 - 1 \times 1 = 17$, $3 \times 3 - 1 \times 2 = 7$. This is the correct number but the wrong sign. So the 3×3 must be negative, which means that the second term must be $^-3$ in one of the brackets.

So the answer is $(3x + 2)(x - 3)$. As there was no common factor the 3s must be in separate brackets.

Multiply out to check.

EXAMPLE 14

Factorise $6x^2 + 3x - 30$.

$3(2x^2 + x - 10)$

The common factor is 3.

Look at the quadratic expression. The signs are different.

First terms are 2 and 1, second terms are 1 and 10, or 10 and 1 or 5 and 2 or 2 and 5.

Try the easiest products first.

Middle term products are $2 \times 5 - 1 \times 2 = 8$, $2 \times 2 - 1 \times 5 = -1$ which is correct but the wrong sign.

So it is $^-2$ in one bracket and the answer is $3(2x + 5)(x - 2)$. Check by multiplying out.

EXAMPLE 15

Factorise $6x^2 - 5x - 4$.

There is no common factor.

The signs are different.

First terms are 6 and 1 or 3 and 2, second terms are 4 and 1 or 1 and 4 or 2 and 2.

Try the easiest products first.

$3 \times 2 - 2 \times 2 = 2$, $3 \times 4 - 2 \times 1 = 10$,
$3 \times 1 - 2 \times 4 = 5$ which is correct.

So the answer is $(3x - 4)(2x + 1)$. Check by multiplying out.

Exam tip

If the sign of c is negative, find differences of products, then put numbers in brackets and lastly signs. Always check by multiplying out.

EXERCISE 2.6A

Factorise the following.

1. $x^2 - x - 6$
2. $x^2 + 3x - 10$
3. $2x^2 + 5x - 3$
4. $3x^2 - 2x - 8$
5. $2x^2 + 9x - 5$
6. $5x^2 - 15x - 50$
7. $4x^2 - 4x - 3$
8. $3x^2 - x - 14$
9. $2x^2 - x - 21$
10. $6x^2 - 17x - 14$

EXERCISE 2.6B

Factorise the following.

1. $x^2 - 3x - 18$
2. $3x^2 + x - 10$
3. $2x^2 - 18$
4. $3x^2 - 11x - 4$
5. $3x^2 + 4x - 15$
6. $5x^2 + 13x - 6$
7. $7x^2 + 10x - 8$
8. $3x^2 - 11x - 20$
9. $2x^2 - 15x - 8$
10. $6x^2 - 13x - 15$

Cancelling fractions

When cancelling fractions it is factors that cancel, never part of factors.

EXAMPLE 16

Simplify $\dfrac{4ab^2}{3c^2} \times \dfrac{9c^2}{2a^2b}$

$\dfrac{4ab^2}{3c^2} \times \dfrac{9c^2}{2a^2b} = \dfrac{6b}{a}$

2, 3, a, b and c^2 all cancel.

EXAMPLE 17

Simplify $\dfrac{x^2 + x}{x^2 - 2x - 3}$

As it stands it cannot be cancelled. First both numerator and denominator must be factorised.

$\dfrac{x^2 + x}{x^2 - 2x - 3} = \dfrac{x(x+1)}{(x-3)(x+1)} = \dfrac{x}{(x-3)}$

$(x + 1)$ cancels.

Exam tip

Errors often occur by cancelling individual terms. Only factors, which can be individual numbers, letters or brackets, can be cancelled.

EXERCISE 2.7A

Simplify

1 $\dfrac{15a^2}{6} \times \dfrac{b^2}{a}$

2 $\dfrac{x^3y^2}{10x} \times \dfrac{15xy}{10y^2}$

3 $\dfrac{2x}{x^2 - 3x}$

4 $\dfrac{3x^2 - 6x}{x^2 + x - 6}$

5 $\dfrac{x^2 - 5x + 4}{x^2 - 2x - 8}$

6 $\dfrac{x^2 - 5x + 6}{x^2 - 9}$

7 $\dfrac{2x^2 + 3x - 5}{4x^2 - 25}$

8 $\dfrac{5x^2 - 80}{2x^2 - 8x}$

EXERCISE 2.7B

Simplify

1 $\dfrac{12abc}{a^2b} \times \dfrac{a^3b}{4c}$

2 $\dfrac{2x^2y^2}{xy} \times \dfrac{3xy^3}{4x^2}$

3 $\dfrac{5x^2 - 20x}{10x^2}$

4 $\dfrac{x^2 + 2x + 1}{x^2 - 1}$

5 $\dfrac{3x^2 + 5x - 2}{x^2 + 7x + 10}$

6 $\dfrac{x^2 - 16}{3x^2 + 12x}$

7 $\dfrac{3x^2 + 3x - 18}{2x^2 - 18}$

8 $\dfrac{3x^2 - x - 4}{5x^2 - 5}$

Key ideas

- When multiplying two brackets, multiply every term in the first bracket by every term in the second bracket.
- When multiplying or dividing algebraic expressions involving powers, add or subtract the indices.
- When finding common factors, make sure you factorise fully.
- When there is no 'x term' in a quadratic equation, the difference of two squares method is likely to be needed.
- When factorising, look first for common factors.
- When factorising $ax^2 + bx + c$ find numbers to fit

  $(px + q)(rx + s)$ that is $pr = a$, $qs = c$ and $ps + qr = b$.

  If c is positive, it is a sum and if c is negative, it is a difference.
- Always check by multiplying out the brackets.
- When cancelling algebraic fractions, factorise if necessary and then cancel factors.

3 Proportion and variation

You should already know

- how to find and use multipliers
- simple algebraic manipulation of formulae and substitution.

You have met proportion before. Look at this example.

EXAMPLE 1

A car uses 12 litres of petrol to travel 100 km. How many litres will it use to travel 250 km?

This is an example of **direct proportion**. As the distance increases, so does the amount of petrol used.

The distance has been increased by multiplying by a factor of $\frac{250}{100} = 2 \cdot 5$.

Increase the amount of petrol by multiplying by the same factor.

Amount of petrol = $12 \times 2 \cdot 5 = 30$ litres.

In this example, if the distance is x km and the amount of petrol is y litres, then y is directly proportional to x. In other words, y varies in the same way as x. If you double x, you also double y. It is possible to express this in symbols:

$$y \propto x,$$

which is read as 'y varies as x' or 'y is proportional to x'.

Now look at this example.

EXAMPLE 2

It takes 4 men 10 days to dig a hole.
How long will it take 20 men?

This is an example of **inverse proportion**. As the number of men increases, the time taken will decrease.

Example 2 cont'd

The number of men has increased by multiplying by a factor of $\frac{20}{4} = 5$.

Decrease the time by dividing by the same factor.

Number of days = $10 \div 5 = 2$.

(Of course, this does assume that there will be room for all the men in the hole!)

This time, if the number of men is x and the number of days is y, then y is inversely proportional to x. In other words,

y varies inversely as x. If you double x, you half y. This is the same as y varies as $\frac{1}{x}$, so in symbols:

$$y \propto \frac{1}{x}.$$

Exam tip

You can easily tell whether the proportion is direct or inverse – in direct, both variables change in the same way, either up or down; in inverse, when one variable goes up, the other will go down.

EXERCISE 3.1A

1 Describe the variation in each of these, using the symbol $\propto$.

  a) The length of tape, y, and the time of the recording, x.

  b) The cost of a train ticket, y, and the length of the journey, x.

  c) The time the journey takes, t, and the speed of the train, s.

  d) The number of pages in a book, p, and the number of words, w.

  e) The probability my ticket wins the raffle, p, and the number of raffle tickets sold, n.

2 Describe the variation shown in each of these tables of values. Use the symbol $\propto$.

a)

x	3	15
y	1	5

b)

x	3	15
y	2	10

c)

x	8	20
y	10	4

d)

x	10	12
y	50	60

e)

x	24	4·8
y	16	3·2

EXERCISE 3.1B

1 Describe the variation in each of these, using the symbol $\propto$.

a) The depth of water in a rectangular tank, d, and the length of time it has been filling, t.

b) The number of buses, b, needed to carry 2000 people and the number of seats on a bus, s.

c) The time a journey takes, t, and the distance covered at a fixed speed, s.

d) The number of ice creams you can buy, c, and the amount of money you have, m.

e) The probability I win the raffle, p, and the number of raffle tickets I buy, n.

2 Describe the variation shown in each of these tables of values. Use the symbol $\propto$.

a)

x	3	15
y	5	1

b)

x	3	15
y	2	10

c)

x	8	20
y	10	25

d)

x	1	0·1
y	50	500

e)

x	16	56
y	6·4	22·4

Variation as a formula

If $y \propto x$ then the same factor is applied to y as was applied to x.

EXAMPLE 3

x	5	15
y	3	9

$y \propto x$ as $5 \times 3 = 15$ for x and $3 \times 3 = 9$ for y.

Now look at the pairs of values of x and y, (5, 3) and (15, 9). In the first $\dfrac{y}{x} = \dfrac{3}{5}$.

But this is also true for the second pair, as $\dfrac{9}{15} = \dfrac{3}{5}$.

Rewriting the equation gives a formula $y = \dfrac{3}{5}x$.

Another way to look at the problem is to graph the equation $y = \dfrac{3}{5}x$.

Check that the points (5,3) and (15,9) are on the graph.

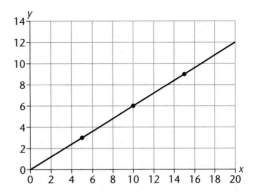

EXAMPLE 4

Find the formula for Example 2.

x	4	20
y	10	2

This time $y \propto \dfrac{1}{x}$.

Write the values of $\dfrac{1}{x}$ in the table.

$\dfrac{1}{x}$	0·25	0·05
y	10	2

So now

$$y \div \frac{1}{x} = 10 \div 0.25$$

Hence $y \div \dfrac{1}{x} = 40$.

This gives the formula $y = \dfrac{40}{x}$.

Check that it works in the table.

There is another way to do this, which you may have spotted already.

In the first table, find the values of $x \times y$. This is 40, so the formula is $xy = 40$, which is equivalent to the first.

Here is the graph of $y = \dfrac{40}{x}$.

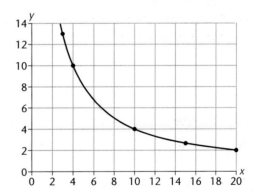

Check that the points (4,10) and (20,2) are on the graph.

Chapter 3   *Proportion and variation*

EXERCISE 3.2A

Find formulae for the variations in Exercise 3.1A, questions 2a) – e). In each case, sketch the graph.

EXERCISE 3.2B

Find formulae for the variations in Exercise 3.1B, questions 2a) – e). In each case, sketch the graph.

Other variation

Sometimes two variables can be related in more complicated ways. In this chapter they will be:

$$y \propto x^2, \ y \propto x^3 \text{ and } y \propto \frac{1}{x^2}$$

EXAMPLE 5

$y \propto x^2$.
If $y = 10$ when $x = 5$, what is y when $x = 15$?

x has been multiplied by $15 \div 5 = 3$, so y will be multiplied by $3^2 = 9$.

$y = 10 \times 9 = 90$.

EXAMPLE 6

$y \propto \frac{1}{x^2}$ This is sometimes called the 'inverse square law', for obvious reasons.

If $y = 10$ when $x = 5$, what is y when $x = 10$?

x has been multiplied by $10 \div 5 = 2$, so y will be divided by $2^2 = 4$.

$y = 10 \div 4 = 2{\cdot}5$.

EXERCISE 3.3A

1 $y \propto x^2$, $y = 3$ when $x = 6$. Find y when $x = 12$.

2 $y \propto x^2$, $y = 9$ when $x = 7{\cdot}5$. Find y when $x = 5$.

3 $y \propto x^3$, $y = 1$ when $x = 3$. Find y when $x = 6$.

4 $y \propto \frac{1}{x^2}$, $y = 4$ when $x = 4$. Find y when $x = 8$.

5 $y \propto x^2$, $y = 5$ when $x = 6$. Find y when $x = 3$.

6 $y \propto \frac{1}{x^2}$, $y = 10$ when $x = 3$. Find y when $x = 12$.

7 $y \propto \frac{1}{x^2}$, $y = 10$ when $x = 4$. Find y when $x = 6$.

8 $y \propto x^3$, $y = 12$ when $x = 5$. Find y when $x = 10$.

9 $y \propto x^3$, $y = 3$ when $x = 10$. Find y when $x = 5$.

10 $y \propto x^2$, $y = 7$ when $x = 8$. Find y when $x = 6$.

Exercise 3.3A cont'd

11 Describe the variation shown in each of these tables of values. Use the symbol ∝.

a)

x	5	25
y	5	125

b)

x	5	10
y	5	1·25

c)

x	5	15
y	5	135

d)

x	5	2·5
y	5	10

e)

x	4	6
y	18	8

EXERCISE 3.3B

1 $y \propto x^3$, $y = 4$ when $x = 5$. Find y when $x = 10$.

2 $y \propto x^2$, $y = 2$ when $x = 2$. Find y when $x = 8$.

3 $y \propto \dfrac{1}{x^2}$, $y = 7$ when $x = 7$. Find y when $x = 14$.

4 $y \propto \dfrac{1}{x}$, $y = 1$ when $x = 1$. Find y when $x = 0.5$.

5 $y \propto x$, $y = 8$ when $x = 3$. Find y when $x = 10.5$.

6 $y \propto \dfrac{1}{x^2}$, $y = 3$ when $x = 1$. Find y when $x = 0.5$.

7 $y \propto x^3$, $y = 14$ when $x = 12$. Find y when $x = 15$.

8 $y \propto x^2$, $y = 2$ when $x = 6.5$. Find y when $x = 19.5$.

9 $y \propto x^2$, $y = 5$ when $x = 0.6$. Find y when $x = 2.4$.

10 $y \propto \dfrac{1}{x^2}$, $y = 9$ when $x = 3$. Find y when $x = 1$.

> **Exam tip**
>
> When finding the variation or the formula, start by deciding if the proportion is direct or inverse. This reduces the number of possibilities to be tried.

11 Describe the variations shown in each of these tables of values. Use the symbol ∝.

a)

x	2	6
y	7	63

b)

x	1	0.25
y	1	4

c)

x	54	21·6
y	33	13·2

d)

x	16	8
y	15	1·875

e)

x	24	48
y	4	1

Formulae

Finding formulae for other variations can be done in a similar way to the previous ones. There is another approach, however.

EXAMPLE 7

Here is the result of Example 5 in a table.

x	5	15
y	10	90

$y \propto x^2$ so try the formula $y = kx^2$, where k is a constant to be found.

Substitute 5 and 10 for x and y.

$10 = k \times 5^2$, giving $k = 0.4$.

Substitute 15 and 90 as a check.

$90 = 0.4 \times 15^2$, which is correct. So $y = 0.4x^2$.

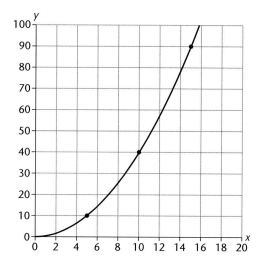

Check that the points from the table are on the graph.

EXAMPLE 8

Here is the result of Example 6 in a table.

x	5	15
y	10	2·5

$y \propto \dfrac{1}{x^2}$, so try the formula $y = \dfrac{k}{x^2}$.

Substitute 5 and 10 for x and y.

$10 = \dfrac{k}{5^2}$, giving $k = 250$.

Substitute 10 and 2·5 as a check.

$2.5 = \dfrac{250}{100}$, which is correct.

So $y = \dfrac{250}{x^2}$.

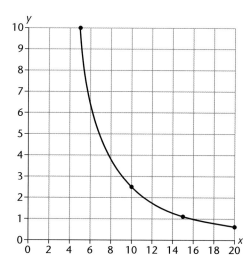

Check that the points from the table are on the graph

Find the formulae for each of the questions in Exercise 3.3A. For each part of question 11, sketch the graph.

Find formulae for each of the questions in Exercise 3.3B. For each part of question 11, sketch the graph.

Key ideas

- The notation '$\propto$' means 'varies as' or 'is proportional to'.
- Direct proportion includes $y \propto x$, $y \propto x^2$ and $y \propto x^3$ and the formulae for these are $y = kx$, $y = kx^2$, $y = kx^3$, respectively.
- Inverse proportion includes $y \propto \dfrac{1}{x}$, $y \propto \dfrac{1}{x^2}$ and the formulae for these are $y = \dfrac{k}{x}$, $y = \dfrac{k}{x^2}$, respectively.

Revision exercise

1 Find the approximate answers to these calculations by rounding each number to one significant figure.

   a) 580×83

   b) $\dfrac{63 \cdot 2}{3 \cdot 8}$

   c) $\dfrac{28 \cdot 3^2}{0 \cdot 48}$

2 Multiply out these brackets.

   a) $(x + 7)(x + 1)$

   b) $(a - 3)(a + 5)$

   c) $(2y - 4)(y + 1)$

   d) $(x - 5)(2x + 1)$

   e) $(4a + b)(a - b)$

3 Simplify:

   a) $2a^2 \times a^3$

   b) $10a^2 \div 2a$

   c) $(a^3)^2 \times a^3 \div a^4$

   d) $12a^2b \times 2a^2b^3$

   e) $6x^2y^2z^2 \div 2xy^2z$

   f) $8a^2b \times \dfrac{3abc}{6ab^2}$

4 Factorise each of these fully.

   a) $3a + 6b - 12c$

   b) $2a + 3ab$

   c) $a^2b - 3ab^2$

   d) $2x^2y - 6xy$

   e) $7abc + 14a^2b$

   f) $9a^2 + 3b^2 - 6c^2$

   g) $5pq - 10$

   h) $2a - 4a^2 + 6a^3$

   i) $100abc - 50ac$

5 Factorise these, where possible.

   a) $x^2 - 49$

   b) $x^2 - 1$

   c) $x^2 + 16$

   d) $x^2 - y^2$

   e) $121 - b^2$

6 Factorise:

   a) $x^2 - 16x + 63$

   b) $2x^2 - 8x - 42$

   c) $3x^2 - 8x + 4$

   d) $2x^2 + x - 15$

   e) $3x^2 - 48$

   f) $2x^2 - 11x - 21$

   g) $6x^2 - 27x - 15$

   h) $5x^2 - 21x + 18$

   i) $8x^2 - 6x - 5$

   j) $6x^2 - 11x - 10$

7 Simplify:

   a) $\dfrac{x^2 + 2x - 3}{4x - 4}$

   b) $\dfrac{x^2 + 2x + 1}{x^2 - x - 2}$

   c) $\dfrac{2x^2 + 2x}{x^2 + 4x + 3}$

   d) $\dfrac{x^2 - 7x + 10}{2x^2 - 11x + 5}$

   e) $\dfrac{x^2 + 8x + 15}{x^2 - 25}$

   f) $\dfrac{3x^2 + 15x + 12}{2x^2 + 7x - 4}$

8 The attractive force between two objects is inversely proportional to the square of their distance apart. The force is 0.24 units when the distance is 15 units.

 a) What will be the force when the distance is 30 units?

 b) Find a formula for the force, f, in terms of the distance apart, d.

9 Complete the table to find the corresponding values of y.

	$y \propto x$	$y \propto x^2$	$y \propto x^3$	$y \propto \frac{1}{x}$	$y \propto \frac{1}{x^2}$
x	2	10	10	10	10	10
y	5					

10 In each part, find the variation, using $\propto$, and find the formula.

a)

x	5	10
y	10	5

b)

x	5	50
y	10	100

c)

x	2	10
y	0.1	0.02

d)

x	2	10
y	0.1	2.5

4 Indices

You should already know

- how to work out numbers with positive and negative indices
- how to use basic rules of indices
- how to give answers to a number of significant figures
- the meaning of the words powers, primes and factors.

Fractional indices

You have already looked at numbers with positive and negative indices and should know

that $\qquad n^4 = n \times n \times n \times n,$ $\qquad n^{-2} = \dfrac{1}{n^2},$ $\qquad \left(\dfrac{a}{b}\right)^{-1} = \dfrac{b}{a}$ $\qquad$ and $\qquad n^0 = 1.$

Suppose $a^b = \sqrt[3]{a}$

Then $a^b \times a^b \times a^b = (\sqrt[3]{a})^3 = a = a^1$ $\quad$ Cube both sides.

$\qquad a^{b+b+b} = a^1$

$\qquad\qquad a^{3b} = a^1$

$\qquad\qquad\quad 3b = 1$ $\qquad\qquad\qquad$ Equate powers.

$\qquad\qquad\qquad b = \dfrac{1}{3}$

Therefore $a^{\frac{1}{3}} = \sqrt[3]{a}$

A similar proof can be given to show that $a^{\frac{1}{2}} = \sqrt{a}$ etc., and also $a^{\frac{1}{n}} = \sqrt[n]{a}$.

Similarly $a^{\frac{3}{2}} = (a^{\frac{1}{2}})^3 = (\sqrt{a})^3$ or $(a^3)^{\frac{1}{2}} = \sqrt{a^3}$

EXAMPLE 1

Write the following in index form, as simply as possible.

a) the cube of n $\qquad$ **b)** $\dfrac{1}{n^3}$ $\qquad$ **c)** $\sqrt[5]{n}$

a) n^3 $\qquad\qquad$ **b)** n^{-3} $\qquad$ **c)** $n^{\frac{1}{5}}$

You will be asked to work out powers of numbers both with and without a calculator.

EXAMPLE 2

Write as whole number or fractions. Do not use a calculator.

a) 3^2 **b)** $16^{\frac{1}{4}}$

c) $343^{\frac{1}{3}}$ **d)** 4^{-2}

e) $\left(\dfrac{1}{3}\right)^{-2}$ **f)** 6^0

g) $125^{\frac{2}{3}}$

a) $3 \times 3 = 9$ **b)** 2 $(2 \times 2 \times 2 \times 2 = 16)$

c) 7 $(7 \times 7 \times 7 = 343)$ **d)** $\dfrac{1}{4^2} = \dfrac{1}{16}$

e) $3^2 = 9$ **f)** 1

g) 25 $((\sqrt[3]{125})^2 = 5^2 = 25)$

> **Exam tip**
>
> If you have to work out the square root of the cube of a number, it is normally easier to find the root first.

Working with a calculator

On your calculator you will see a key labelled x^y or y^x, which will enable you to work out numbers like $3 \cdot 1^5$ etc. If you are not sure how to use it, practise with something simple like 2^4 which you can work out as 16. Try '2' 'x^y' '4' '='.

Similarly, you will have a key labelled $x^{\frac{1}{y}}$ or similar which may be 'SHIFT' 'x^y'. This will enable you to work out numbers like $2 \cdot 5^{\frac{1}{4}}$. Again practise with calculations you can work out in your head to check you are correct. You could do Example 2 again, this time using a calculator to check how your calculator works.

EXAMPLE 3

Use a calculator to work out the following. Give your answers exactly or to 5 significant figures.

a) $3 \cdot 5^4$ **b)** $2 \cdot 4^6$ **c)** $1 \cdot 03^{-3}$

d) $2 \cdot 15^{\frac{1}{4}}$ **e)** $3125^{\frac{4}{5}}$

a) $150 \cdot 0625 = 150 \cdot 06$ **b)** $191 \cdot 1029 = 191 \cdot 10$

c) $(0 \cdot 9708737)^3 = 0 \cdot 9151416 = 0 \cdot 91514$

> Note that when the index is negative, it may be best first to find the reciprocal and then raise to a positive index.

d) $1 \cdot 210903 = 1 \cdot 2109$ **e)** 625

> **Exam tip**
>
> When finding a root by calculator it is easy to make a mistake. It is very helpful to check by working backwards. For example, in **d)** above $2 \cdot 15^{\frac{1}{4}} = 1 \cdot 210903$. Check $1 \cdot 210903^4$ $= 2 \cdot 14999 \simeq 2 \cdot 15$ so it checks.

EXERCISE 4.1A

1   Write in index form: **a)** cube root of n **b)** the reciprocal of n^3 **c)** $5\sqrt{n^2}$

Work out the following without using a calculator.

Give the answers as whole numbers or fractions.

2   **a)** 4^{-1} **b)** $4^{\frac{1}{2}}$ **c)** 4^0 **d)** 4^{-2} **e)** $4^{\frac{3}{2}}$

3   **a)** $8^{\frac{1}{3}}$ **b)** 8^{-1} **c)** $8^{\frac{4}{3}}$ **d)** $\left(\frac{1}{8}\right)^{-2}$ **e)** 8^1

4   **a)** $64^{\frac{1}{2}}$ **b)** $64^{-\frac{1}{3}}$ **c)** 64^0 **d)** $64^{\frac{2}{3}}$ **e)** $64^{\frac{5}{6}}$

5   **a)** $2^2 \times 9^{\frac{1}{2}}$ **b)** $2^5 \times 8^{\frac{1}{3}}$ **c)** $81^{\frac{1}{4}} \times 3^{-2}$ **d)** $9^{\frac{1}{2}} \times 6^2 \times 4^{-1}$

6   **a)** $2^2 + 3^0 + 16^{\frac{1}{2}}$ **b)** $\left(\frac{3}{4}\right)^{-2} \times 27^{\frac{2}{3}}$ **c)** $4^2 \div 9^{\frac{1}{2}}$ **d)** $4^2 - 8^{\frac{1}{3}} + 9^0$

Use a calculator for these questions. Give the answers exactly or correct to 5 s.f.

7   **a)** $1 \cdot 14^5$ **b)** $2 \cdot 79^3$ **c)** $1 \cdot 005^9$ **d)** $4 \cdot 1^{-4}$

8   **a)** $923521^{\frac{1}{4}}$ **b)** $1 \cdot 051^{\frac{1}{5}}$ **c)** $21^{\frac{1}{7}}$ **d)** $6 \cdot 45^{\frac{2}{5}}$

9   **a)** $100 \times 1 \cdot 02^3$ **b)** $1 \cdot 6^5 \times 2 \cdot 1^{\frac{1}{3}}$ **c)** $(10^5 \times 4 \cdot 1)^{\frac{1}{4}}$

10 **a)** $1 \cdot 9^4 - 2 \cdot 1^3$ **b)** $1 \cdot 9^{\frac{1}{4}} + 0 \cdot 97^{\frac{1}{5}}$ **c)** $14^3 - 196^{\frac{3}{2}}$

EXERCISE 4.1B

1   Write in index form: **a)** $\sqrt[4]{n}$ **b)** the reciprocal of $\left(\frac{1}{n}\right)^4$ **c)** $\sqrt[3]{n^5}$

Work out the following without using a calculator.

Give the answers as whole numbers or fractions.

2   **a)** 9^{-1} **b)** $9^{\frac{1}{2}}$ **c)** 9^0 **d)** 9^{-2} **e)** $9^{\frac{3}{2}}$

3   **a)** $27^{\frac{1}{3}}$ **b)** $27^{\frac{4}{3}}$ **c)** 27^{-1} **d)** $\left(\frac{1}{27}\right)^{-\frac{1}{3}}$ **e)** 27^0

4   **a)** $16^{\frac{1}{2}}$ **b)** $16^{-\frac{1}{4}}$ **c)** 16^0 **d)** $16^{\frac{3}{2}}$ **e)** $16^{\frac{7}{4}}$

5   **a)** $25^{\frac{3}{2}}$ **b)** $36^{\frac{1}{2}}$ **c)** $125^{\frac{2}{3}} \times 8^{\frac{2}{3}}$ **d)** $49^{\frac{3}{2}} \times 81^{-\frac{1}{4}}$

6   **a)** $5^{-2} \times 10^5 \times 16^{-\frac{1}{2}}$ **b)** $\left(\frac{4}{5}\right)^2 \times 128^{-\frac{3}{7}}$ **c)** $5^3 - 25^{\frac{1}{2}} - \left(\frac{2}{5}\right)^{-2}$ **d)** $125^{\frac{1}{3}} - 121^{\frac{1}{2}} + 216^{\frac{1}{3}}$

Use a calculator for these questions. Give the answers exactly or correct to 5 s.f.

7   **a)** $3 \cdot 25^4$ **b)** $0 \cdot 46^5$ **c)** $1 \cdot 01^7$ **d)** $2 \cdot 91^{-3}$

8   **a)** $14641^{\frac{1}{4}}$ **b)** $14120^{\frac{1}{5}}$ **c)** $9^{\frac{1}{9}}$ **d)** $1024^{\frac{2}{5}}$

9   **a)** $4^3 + 3^4$ **b)** $1 \cdot 6^4 \times 1 \cdot 7^{\frac{1}{4}}$ **c)** $1^5 \times 4 \cdot 1^{\frac{1}{4}}$

10 **a)** $5 \cdot 27^5 - 3 \cdot 49^5$ **b)** $4^{\frac{3}{4}} + 5^{\frac{2}{5}}$ **c)** $216^{\frac{4}{3}} \times 9^{-\frac{2}{3}}$

Without the use of a calculator, match each indexed number to a whole number.

2^3 $\left(\frac{1}{2}\right)^{-2}$ $1000^{\frac{2}{3}}$ $64^{\frac{1}{2}}$ 10^2 $64^{\frac{1}{3}}$ $(0.05)^{-1}$ $8^{\frac{2}{3}}$ $4^{\frac{1}{2}}$ $64^{\frac{2}{3}}$ $64^{\frac{1}{6}}$ 7^0 $36^{\frac{1}{2}}$ $64^{\frac{5}{6}}$

0   1   2   4   6   8   9   10   16   20   32   64   100   200

Using rules of indices with numbers and letters

You have already learnt the rules for indices:

$a^n \times a^m = a^{n+m}$, $a^n \div a^m = a^{n-m}$, $(a^n)^m = a^{n \times m}$.

These can be used with either numbers or letters.

EXAMPLE 4

Write as a single power of 2 where possible:

a) $2\sqrt{2}$ **b)** $(^3\sqrt{2})^2$

c) $2^3 \div 2^{\frac{1}{2}}$ **d)** $2^3 + 2^4$

e) $8^{\frac{3}{4}}$ **f)** $2^3 \times 4^{\frac{3}{2}}$

g) $2^n \times 4^3$

a) $2^1 \times 2^{\frac{1}{2}} = 2^{\frac{3}{2}}$

b) $(2^{\frac{1}{3}})^2 = 2^{\frac{2}{3}}$

c) $2^{3-\frac{1}{2}} = 2^{2\frac{1}{2}} = 2^{\frac{5}{2}}$

d) $2^3 + 2^4$ These powers cannot be added.

e) $(2^3)^{\frac{3}{4}} = 2^{3 \times \frac{3}{4}} = 2^{\frac{9}{4}}$

f) $2^3 \times (2^2)^{\frac{3}{2}} = 2^3 \times 2^3 = 2^6$

g) $2^n \times (2^2)^3 = 2^n \times 2^6 = 2^{n+6}$

Exam tip

The most common mistakes are trying to add or subtract a^x and a^y which cannot be done.

EXAMPLE 5

Write 132 as a product of its prime factors. Use indices where possible.

$132 = 2 \times 66 = 2 \times 2 \times 33 = 2 \times 2 \times 3 \times 11$
$= 2^2 \times 3 \times 11$.

EXERCISE 4.2A

1   Write as powers of 3 as simply as possible:
   a) 27
   b) $\frac{1}{3}$
   c) $3 \times \sqrt{3}$
   d) $81^{\frac{3}{2}}$
   e) $3^4 \times 9^{-1}$
   f) $9^n \times 27^{3n}$.

2   Write as powers of 5 as simply as possible:
   a) 625
   b) $25^{\frac{-1}{2}} \times 5^3$
   c) 0·2
   d) $125^{\frac{3}{2}} \times 5^{-3} \div 25^2$
   e) $5^4 - 5^3$
   f) $25^{3n} \times 125^{\frac{n}{3}}$

3   Write as powers of 2 and 3 as simply as possible:
   a) 24
   b) $6^2 \times 4^2$
   c) $18^{\frac{1}{3}}$
   d) $\frac{4}{9}$
   e) $13\frac{1}{2}$
   f) 12^{2n}

4   Write as product of the prime factors. Use indices where possible.
   a) 75
   b) 144
   c) 300
   d) 324

EXERCISE 4.2B

1   Write as powers of 2 as simply as possible:
   a) 32
   b) $8^{\frac{2}{3}}$
   c) $2 \times \sqrt[3]{64}$
   d) 0·25
   e) $2^{2n} \times 4^{\frac{n}{2}}$
   f) $2^{3n} \times 16^{-2}$

2   Write each as a power of a prime number as simply as possible.
   a) 343
   b) $25^{\frac{1}{6}}$
   c) $16^{\frac{1}{2}} \times 64^{\frac{-2}{3}}$
   d) $27^2 \div 81^{\frac{3}{2}}$
   e) $2^5 + 2^2$
   f) $9^{2n} \times 3^{-2n}$

3   Write as a product of the prime factors. Use indices where possible.
   a) 36
   b) 96
   c) 60
   d) 392

4   Write as products of prime numbers:
   a) 15^3
   b) $12^{\frac{1}{2}} \times 9^{\frac{-1}{4}}$
   c) 40^n
   d) $20^{2n} \times 100^n$

Key ideas

● The rules of indices are $a^n \times a^m = a^{m+n}$,
   $a^n \div a^m = a^{n-m}$, $(a^n)^m = a^{n \times m}$, $a^0 = 1$,
   $a^{-n} = \frac{1}{a^n}$, $a^{\frac{1}{n}} = \sqrt[n]{a}$

● The use of the keys x^y and $x^{\frac{1}{y}}$ needs to be learnt for your calculator.

● When working out $(3a^2)^2$ be careful that the number part is 3×3 not 3×2.

5 Rearranging formulae

You should already know

- how to factorise simple expressions
- how to rearrange simple formulae
- how to expand brackets and manipulate simple algebraic expressions.

All the formulae that have been covered previously contained the new subject only once and always as part of the numerator. This is now extended.

EXAMPLE 1

Rearrange the formula $a = x + \dfrac{cx}{d}$ to make x the subject.

$ad = dx + cx$	Multiply through by d.
$dx + cx = ad$	Rearrange to get all terms involving x on the left-hand side.
$x(d + c) = ad$	Factorise.
$x = \dfrac{ad}{d + c}$	Divide by $(d + c)$.

EXAMPLE 2

Rearrange the formula $a = \dfrac{1}{p} + \dfrac{1}{q}$ to make p the subject.

$apq = q + p$	Multiply through by pq.
$apq - p = q$	Collect all terms in p to the left-hand side.
$p(aq - 1) = q$	Factorise.
$p = \dfrac{q}{aq - 1}$	Divide by $(aq - 1)$.

EXAMPLE 3

Rearrange the formula $a = b + \dfrac{c}{1 + p}$ to make p the subject.

$a(1 + p) = b(1 + p) + c$	Multiply through by $(1 + p)$.
$a + ap = b + bp + c$	Expand brackets.
$ap - bp = b + c - a$	Collect all terms in p to the left-hand side.
$p(a - b) = b + c - a$	Factorise.
$p = \dfrac{b + c - a}{a - b}$	Divide by $(a - b)$

EXAMPLE 4

Rearrange the formula $s = b + \sqrt{\dfrac{t}{p}}$ to make t the subject.

$s = b + \sqrt{\dfrac{t}{p}}$	If a root or power is involved, rearrange to get that by itself.
$(s - b)^2 = \dfrac{t}{p}$	Square both sides.
$\dfrac{t}{p} = (s - b)^2$	Get all terms in t to the left-hand side.
$t = p(s - b)^2$	Multiply through by p.
	In this case there is no need to expand $(s - b)^2$.

EXERCISE 5.1A

For each question, rearrange to make the letter in square brackets the subject.

1 $s = at + 2bt$ [t]

2 $P = t - \dfrac{at}{b}$ [t]

3 $s - 2ax = b(x - s)$ [x]

4 $a = \dfrac{t}{b} - st$ [t]

5 $a = \dfrac{1}{b + c}$ [c]

6 $a = b + \dfrac{c}{d + 1}$ [d]

7 $a = b + c^2$ [c]

8 $A = P + \dfrac{PRT}{100}$ [P]

9 $\dfrac{a}{x + 1} = \dfrac{b}{2x - 1}$ [x]

10 $T = 2\pi\sqrt{\dfrac{L}{g}}$ [L]

EXERCISE 5.1B

For each question, rearrange to make the letter in square brackets the subject.

1 $s = ab - bc$ [b]

2 $s = \dfrac{1}{a} + b$ [a]

3 $3(a + y) = by + 7$ [y]

4 $2(a - 1) = b(1 - 2a)$ [a]

5 $\dfrac{a}{b} - 2a = b$ [a]

6 $m = \dfrac{100(a - b)}{b}$ [b]

7 $\dfrac{a}{p} = \dfrac{1}{1 + p}$ [p]

8 $a = \dfrac{1}{1 + x} - b$ [x]

9 $s = 2r^2 - 1$ [r]

10 $s = \dfrac{uv}{u + v}$ [v]

Key idea

● When rearranging a formula, first multiply through by any common denominator, then collect all terms involving the new subject on the left-hand side, then factorise and divide by the factor.

Chapter 5 *Rearranging formulae*

6 Arcs, sectors and volumes

You should already know

- how to find the circumference and area of circles
- how to find the volume of a prism
- how to rearrange formulae
- how to use Pythagoras' theorem and trigonometry.

Arcs and sectors

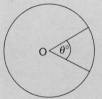

A sector is a fraction of a circle. It is $\dfrac{\theta}{360}$ ths of the circle, where $\theta°$ is the sector angle at the centre of the circle.

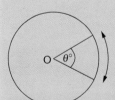

Arc length $= \dfrac{\theta}{360} \times$ circumference $= \dfrac{\theta}{360} \times 2\pi r$

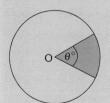

Sector area $= \dfrac{\theta}{360} \times$ area of circle $= \dfrac{\theta}{360} \times \pi r^2$

EXAMPLE 1

Calculate the arc length and area of this sector.

Arc length $= \dfrac{\theta}{360} \times 2\pi r$

$\qquad = \dfrac{37}{360} \times 2\pi \times 5 \cdot 6$

$\qquad = 3 \cdot 62$ cm to 3 significant figures.

Sector area $= \dfrac{\theta}{360} \times \pi r^2$

$\qquad = \dfrac{37}{360} \times \pi \times 5 \cdot 6^2$

$\qquad = 10 \cdot 1$ cm^2 to 3 significant figures.

5·6 cm
37° O

EXAMPLE 2

Calculate the sector angle of a sector with arc length 6·2 cm in a circle with radius 7·5 cm.

Arc length $= \dfrac{\theta}{360} \times 2\pi r$

$6 \cdot 2 = \dfrac{\theta}{360} \times 2\pi \times 7 \cdot 5$

$\theta = \dfrac{6 \cdot 2 \times 360}{2\pi \times 7 \cdot 5}$

$\qquad = 47 \cdot 4°$ to 1 decimal place.

6·2 cm
θ 7·5 cm

Exam tip

You can rearrange the formula before you substitute, if you prefer.

EXAMPLE 3

A sector makes an angle of 54° at the centre of a circle. The area of the sector is 15 cm^2.

Calculate the radius of the circle.

Sector area $= \dfrac{\theta}{360} \times \pi r^2$

$15 = \dfrac{54}{360} \times \pi r^2$

$r^2 = \dfrac{15 \times 360}{54 \times \pi}$

$\qquad = 31 \cdot 83 \ldots$

$r = \sqrt{31 \cdot 83 \ldots}$

$\qquad = 5 \cdot 64$ cm to 3 significant figures.

EXERCISE 6.1A

1 Calculate the arc length of these sectors. Give your answers to 3 significant figures.

a)

b)

c)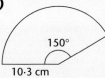

d)

e)

2 Calculate the areas of the sectors in question 1. Give your answer to 3 significant figures.

3 Calculate the perimeters of these sectors. Give your answers to 3 significant figures.

a)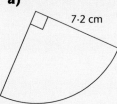

b)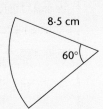

c)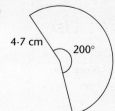

4 Calculate the sector angle in these sectors. Give your answers to the nearest degree.

a)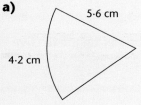

b)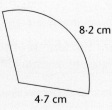

c)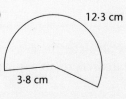

d)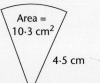

e)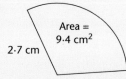

Exercise 6.1A cont'd

5 Calculate the radius of these sectors.

a)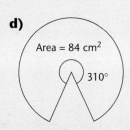

70° 6 cm

b)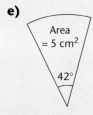

8·9 cm 150°

c)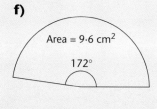

19 cm 225°

d)

Area = 84 cm² 310°

e)

Area = 5 cm² 42°

f)

Area = 9·6 cm² 172°

6 A motif consists of two separate sectors of a circle, each with angle 35° and radius 32 mm. They are to be painted blue with a thin yellow border.

Calculate the areas of blue and the length of the yellow border required. Give your answers to an appropriate degree of accuracy.

EXERCISE 6.1B

1 Calculate the arc length of these sectors. Give your answers to 3 significant figures.

a)
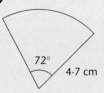

b)

c)
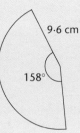

d)
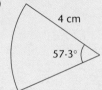

e)
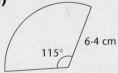

2 Calculate the areas of the sectors in question 1. Give your answers to 3 significant figures.

3 Calculate the perimeters of these sectors. Give your answers to 3 significant figures.

a)
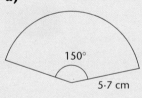

b)
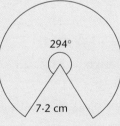

c)

Exercise 6.1B cont'd

4 Calculate the sector angle in these factors. Give your answer to the nearest degree.

a)

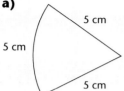

5 cm

5 cm

5 cm

b)

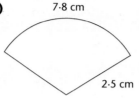

7·8 cm

2·5 cm

c)

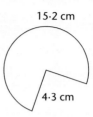

15·2 cm

4·3 cm

d)

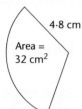

4·8 cm

Area = 32 cm²

e)

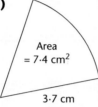

Area = 7·4 cm²

3·7 cm

f)

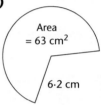

Area = 63 cm²

6·2 cm

5 Calculate the radius of these sectors.

a)

42°

9·8 cm

b)

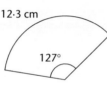

12·3 cm

127°

c)

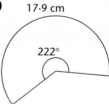

17·9 cm

222°

d) Sector area = 19·7 cm², sector angle = 52°.

e) Sector area = 2·7 cm², sector angle = 136°.

f) Sector area = 6·2 m², sector angle = 218°.

6 A lawn is in the shape of a sector of a circle with angle 63° and radius of 25 m.

a) The owner wants to spread fertiliser on the lawn. Calculate the area needed to be covered.

b) The owner wants to put edging around the lawn. Calculate the length of edging needed.

Give your answers to an appropriate degree of accuracy.

Volumes

In an earlier chapter you learnt how to find the volume of a prism – a shape where the cross-section stays the same throughout its length. Therefore:

volume of a prism = area of cross-section × length.

For a shape whose cross-section, though similar, decreases to a point, its volume is given by

volume = $\frac{1}{3}$ area of cross-section at base × height.

The particular shapes of this type that you need to know are the pyramid, with a square or rectangular cross-section, and the cone. So, for a cone with base radius r and height h,

volume = $\frac{1}{3}\pi r^2 h$

You also need to know about a different type of 3D shape – the sphere.

volume of a sphere of radius $r = \frac{4}{3}\pi r^3$.

EXAMPLE 4

Find the volume of the cone.

a) Volume = $\frac{1}{3}\pi \times 3^2 \times 4$

$\qquad = 12\pi$

$\qquad = 37\cdot7\,\text{cm}^3$.

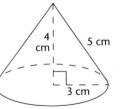

41

EXERCISE 6.2A

1 Calculate the volumes of these pyramids. Their bases are squares or rectangles.

a)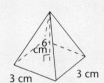

b)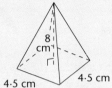

c)

2 Calculate the volumes of these cones.

a)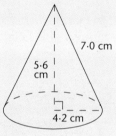

b)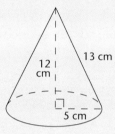

c)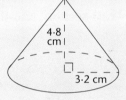

3 Find the volume of a sphere of radius:

   a) 5 cm   **b)** 6·2 cm   **c)** 2 mm.

4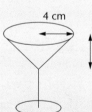

Calculate the capacity of this glass. Give your answer in millilitres.

5 How many ball bearings of radius 0·3 cm can be made from 10 cm³ of metal when it is melted?

6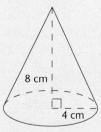

A sphere has the same volume as this cone. Calculate the radius of the sphere.

EXERCISE 6.2B

1 Calculate the volumes of these pyramids. Their bases are squares or rectangles.

a)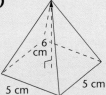

6 cm

5 cm   5 cm

b)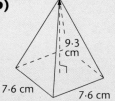

9.3 cm

7.6 cm   7.6 cm

c)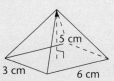

5 cm

3 cm   6 cm

2 Calculate the volumes of these cones.

a)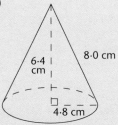

6.4 cm   8.0 cm

4.8 cm

b)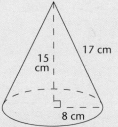

15 cm   17 cm

8 cm

c)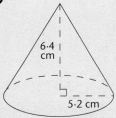

6.4 cm

5.2 cm

3 Find the volume of a sphere of radius:

a) 3 cm   **b)** 4.7 cm   **c)** 5 mm.

4

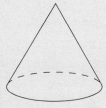

This glass paperweight in the shape of a cone has a volume of 75 cm³. Its base radius is 3 cm. Calculate its height.

5 How many glass marbles of radius 7 mm can be made from 100 cm³ of glass?

6  A plastic pipe is a cylinder 2 m long. The internal and external radii of the pipe are 5 mm and 6 mm. Calculate the volume of plastic in the pipe.

7 A solid cone and a solid cylinder both have base radius 6 cm. The height of the cylinder is 4 cm. The cone and the cylinder each have the same volume. Find the height of the cone.

Key ideas

● Arc length $= \dfrac{\theta}{360} \times 2\pi r$.

● Sector area $= \dfrac{\theta}{360} \times \pi r^2$.

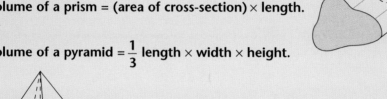

● Volume of a prism = (area of cross-section) $\times$ length.

● Volume of a pyramid $= \dfrac{1}{3}$ length $\times$ width $\times$ height.

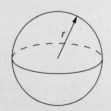

● Volume of a sphere $= \dfrac{4}{3} \pi r^3$.

● Volume of a cone $= \dfrac{1}{3} \pi r^2 h$.

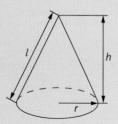

7 Upper and lower bounds

You should already know

- when a measurement is given, e.g. 18·6 seconds to 1 decimal place, its value lies between 18·55 and 18·65 seconds.

Reminder: If $t = 18 \cdot 6$ s to 1 decimal place, then

$$18 \cdot 55 \leq t < 18 \cdot 65$$

This is called the lower bound

This is called the upper bound.

Sums and differences of measurements

Consider two kitchen units, of width 300 mm and 500 mm correct to the nearest mm.

Exam tip

Many people are confused about the upper bound. The convention is that the lower bound is contained in the interval – the upper bound would be in the next higher interval. The measured value 18·6 could be as near 18·65 as you like – 18·6499 or 18·649999 for instance, but the upper bound is 18·65, not even 18·649.

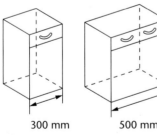

	300 mm	500 mm
Lower bound of width:	299·5 mm	499·5 mm
Upper bound of width:	300·5 mm	500·5 mm

If the units are put next to each other then

the lower bound of w, their joint width = 299·5 + 499·5 = 799 mm

the upper bound of w, their joint width = 300·5 + 500·5 = 801 mm

so $799 \leq w < 801$.

To find the lower bound of a sum, add the corresponding lower bounds.
Similarly, to find the upper bound of a sum, add the upper bounds.

45

However, if we want to consider the difference between the widths of the kitchen units, subtracting a smaller number makes the difference bigger:

upper bound of the difference in their widths = 500·5 − 299·5 = 201 mm

lower bound of the difference in their widths = 499·5 − 300·5 = 199 mm.

> To find the upper bound of a difference, subtract the lower bound of the smaller from the upper bound of the larger.

EXAMPLE 1

A piece of red ribbon is 35·2 cm to the nearest mm. A piece of blue ribbon is 12·6 cm to the nearest mm.

a) What is the minimum length of the two pieces of ribbon laid end to end?

b) What is the lower bound of the difference in the lengths of the two pieces?

For red piece: LB = 35·15 cm   UB = 35·25 cm.

For blue piece: LB = 12·55 cm   UB = 12·65 cm.

a)

Minimum total length = sum of lower bounds

$$= 35·15 + 12·55 = 37 \text{ cm.}$$

b)

Lower bound of difference in lengths = LB of greater piece − UB of smaller piece

$$= 35·15 − 12·65 = 22·5 \text{ cm.}$$

EXERCISE 7.1A

1 Calculate the upper bounds of the sums of these measurements.

   a) 29·7 s and 31·4 s (both to 3 significant figures)

   b) 11·04 s and 13·46 s (both to the nearest hundredth of a second)

   c) 6·42 m and 5·97 m (both to the nearest cm)

   d) 1·248 kg and 0·498 kg (both to the nearest g).

2 Find the lower bounds of the sums of the measurements in question 1.

3 Find the upper bounds of the differences of these measurements:

   a) 947 g and 1650 g (to the nearest g)

   b) 16·4 cm and 9·8 cm (to the nearest mm)

   c) 1650 g and 870 g (to the nearest 10 g)

   d) 24·1 s and 19·8 s (to the nearest 0·1 s).

4 Calculate the lower bounds of the differences between the measurements in question 3.

5 A piece of paper 21·0 cm long is taped side by side to another piece 29·7 cm long, both measurements given to the nearest mm. What is the upper bound of the total length?

EXERCISE 7.1B

1 Calculate the upper bounds of the sums of these measurements:

   a) 86 mm and 98 mm, to the nearest mm

   b) 11·042 kg and 1·695 kg, to the nearest g

   c) 78·5 cm and 69·7 cm, to 3 significant figures

   d) 46·03 s and 59·82 s, to the nearest $\frac{1}{100}$ s.

2 Find the lower bounds of the sums of the measurements in question 1.

3 Find the upper bounds of the differences of these measurements:

   a) 14·86 s and 15·01 s to the nearest $\frac{1}{100}$ s.

   b) 493 m and 568 m to the nearest m

   c) 12 700 m and 3800 m to the nearest 100 m

   d) 1·824 g and 1·687 g to the nearest mg.

4 Calculate the lower bounds of the differences between the measurements in question 3.

5 Two stages of a relay race are run in times of 14·07 s and 15·12 s to the nearest 0·01 s.

   a) Calculate the upper bound of the total time for these two stages.

   b) Calculate the upper bound of the difference between the times for these two stages.

Multiplying and dividing measurements

Consider a piece of A4 paper whose measurements are given as 21·0 cm and 29·7 cm to the nearest mm. What are the upper and lower bounds of the area of the piece of paper?

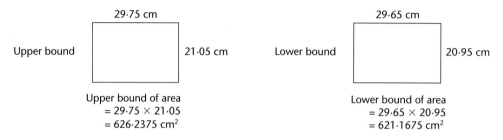

Upper bound

29·75 cm

21·05 cm

Upper bound of area
= 29·75 × 21·05
= 626·2375 cm²

Lower bound

29·65 cm

20·95 cm

Lower bound of area
= 29·65 × 20·95
= 621·1675 cm²

When multiplying:

- multiply the upper bounds to find the upper bound
- multiply the lower bounds to find the lower bound.

When dividing, however, the situation is different.

Dividing by a larger number makes the answer smaller.

When dividing:

- to find the upper bound, divide the upper bound by the lower bound
- to find the lower bound, divide the lower bound by the upper bound.

EXAMPLE 2

Pete cycles 14·2 km (to 3 significant figures) in a time of 46 minutes to the nearest minute. What is the upper bound of his average speed in km/h? Give your answer to 3 significant figures.

$$\text{Upper bound of speed} = \frac{\text{upper bound of distance}}{\text{lower bound of time}}$$

$$= \frac{14\cdot25}{45\cdot5} \text{ km/minute}$$

$$= \frac{14\cdot25}{45\cdot5} \times 60 \text{ km/h}$$

$$= 18\cdot8 \text{ km/h to 3 significant figures.}$$

Exam tip

To find the upper bound of any combined measurement, work out which of the upper and lower bounds of the given measurements you need to use to give you the greatest result. If you aren't sure, experiment!

EXERCISE 7.2A

1 One make of car has a length of 3·2 m, to the nearest 0·1 m. A car transporter of length 16·1 m, to the nearest 0·1 m, wishes to carry 5 cars. Will the 5 cars always fit onto the transporter?

2 Find the upper bound of the floor areas of rectangular rooms with these measurements:

   a) 3·8 m by 4·2 m to 2 significant figures

   b) 5·26 m by 3·89 m to the nearest cm

   c) 8·42 m by 6·75m to 3 significant figures

   d) 7·6 m by 5·2 m to nearest 10 cm.

3 Find the lower bounds of the floor areas of the rooms in question 2.

4 Find the lower bound of the average speeds for these measured times and distances. Give your answers to 3 significant figures.

   a) 6·4 cm in 1·2 s   **b)** 12·4 m in 9·8 s   **c)** 106 m in 10·0 s

5 Calculate the upper bounds of the speeds for the measurements in question 4, giving your answers to 3 significant figures.

6 The mass of an object is given as 1·657 kg to the nearest gram. Its volume is 72·5 cm^3 to 3 significant figures. Find the upper bound of its density. Give your answer to 4 significant figures.

EXERCISE 7.2B

1 Pencils have a width of 8 mm, to the nearest mm. What is the smallest width of a pencil box that can hold 10 pencils side by side?

2 Calculate the upper bounds of the distance travelled for these given speeds and times:

   a) 92·4 cm/s for 12·3 s

   b) 1·54 m/s for 8·2 s

   c) 57 km/h for 2·5 hours

   d) 5·61 m/s for 2·08 s.

3 Calculate the lower bounds of the distances for the data in question 2.

4 Calculate, to 3 significant figures, the minimum width of a rectangle, given the following data:

   a) area = 210 cm^2 to 3 significant figures, length = 17·89 cm to the nearest mm

   b) area = 210 cm^2 to 2 significant figures, length = 19·2 cm to the nearest mm

   c) area = 615 cm^2 to 3 significant figures, length = 30·0 cm to the nearest mm.

Exercise 7.2B cont'd

5 Calculate, to 3 significant figures, the upper bound of the height of the following cuboids:

a) volume = 72 cm^3,
length = 6·2 cm,
width = 4·7 cm
(all to 2 significant figures)

b) volume = 985 cm^3,
length = 17·0 cm,
width = 11·3 cm
(all to 3 significant figures)

c) volume = 84 m^3,
length = 6·2 m,
width = 3·8 m
(all to 2 significant figures)

6 The population of a town is 108 000 to the nearest 1000. Its area is given as 129 square miles. Calculate the upper and lower bounds of its population density, giving your answers to 3 significant figures.

Key ideas

● To find the upper bound of any combined measurement, work out which of the upper and lower bounds of the given measurements you need to use to give you the greatest result. To find the lower bound, you need the smallest result.

Here is a summary of how this works in practice.

● To find the lower bound of a sum, add the corresponding lower bounds. Similarly, to find the upper bound of a sum, add the upper bounds.

● To find the upper bound of a difference, subtract the lower bound of the smaller from the upper bound of the larger.

● The find the lower bound of a difference, subtract the upper bound of the smaller from the lower bound of the larger.

● When multiplying:

multiply the upper bounds to find the upper bound

multiply the lower bounds to find the lower bound.

● When dividing:

to find the upper bound, divide the upper bound by the lower bound

to find the lower bound, divide the lower bound by the upper bound.

Revision exercise

1 Write in index form as simply as possible:

   a) the reciprocal of n

   b) the cube root of m

   c) The square roof of $\dfrac{1}{n}$

2 Write as whole numbers or fractions. Do not use a calculator.

   a) 4^{-1} **b)** 5^0 **c)** $25^{\frac{1}{2}}$

   d) 2^4 **e)** $8^{\frac{2}{3}}$ **f)** $125^{-\frac{2}{3}}$

   g) $\left(\dfrac{1}{8}\right)^{-\frac{1}{3}}$ **h)** $64^{\frac{5}{6}}$ **i)** $\left(\dfrac{4}{9}\right)^{-\frac{1}{2}}$

   j) $\left(\dfrac{1}{12}\right)^{-2}$

3 Write as whole numbers or fractions. Do not use a calculator.

   a) $8^0 \times 25^2$ **b)** $4^2 \times 25^{\frac{1}{2}}$

   c) $12^2 \times 4^{-2}$ **d)** $6^3 \div 9^{\frac{3}{2}}$

   e) $5^2 - 4^3 + 3^4$ **f)** $25^{\frac{3}{2}} \times 64^{\frac{1}{3}}$

   g) $14^2 \times 49^{-1}$ **h)** $\left(\dfrac{4}{5}\right)^{-2} \times \left(\dfrac{16}{9}\right)^{\frac{1}{2}}$

4 Use a calculator for these questions. Give the answers exactly or to 5 significant figures.

   a) $1{\cdot}43^3$ **b)** $0{\cdot}87^5$ **c)** 2^{12} **d)** $7{\cdot}9^{-4}$

5 Use a calculator for these questions. Give the answers exactly or to 5 significant figures.

   a) $59\,049^{\frac{1}{5}}$ **b)** $7{\cdot}9^{\frac{1}{4}}$

   c) $4000^{\frac{1}{6}}$ **d)** $32\,768^{\frac{3}{5}}$

6 Write as a power of a prime number as simply as possible.

   a) 128 **b)** 27^2 **c)** $49^{\frac{1}{3}}$

   d) $9^2 \div 81^{-\frac{1}{2}}$ **e)** $2^n \times 32^{n+1}$

7 Write as a product of the prime factors. Use indices where possible.

   a) 40 **b)** 90

   c) 136 **d)** 588

8 Write as powers of primes as simply as possible.

   a) 15^2 **b)** $40^{\frac{1}{3}}$

   c) $14^3 \times 56^{\frac{1}{3}}$ **d)** $71^{\frac{3}{2}} \times 24^{-\frac{1}{2}}$

9 Rearrange the formulae to make a the subject.

   a) $3a + 5c = 2b - 5a$

   b) $3(a + 2b) = 2b + 5a$

10 Rearrange the formulae to make p the subject.

   a) $p + q = 2(q - 3p)$ **b)** $t = \dfrac{2(p - 1)}{p}$

   c) $\dfrac{1}{p} = \dfrac{1}{q} + \dfrac{1}{s}$ **d)** $T = p + \dfrac{2p}{q}$

   e) $\dfrac{1}{2p - 1} = \dfrac{2a}{p + 1}$ **f)** $p^2 + 4a = 2b$

11 Calculate

   a) the arc length and

   b) the sector area

   of a sector with angle 75° in a circle of radius 6·5 cm.

12 Calculate the sector angle of these sectors.

   a)

   7·2 cm

   8·3 cm

   b)

   4·8 cm   Area = 29 cm²

13 Calculate the radii of the circles with these sectors.

a)
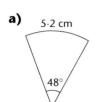
5·2 cm
48°

b)
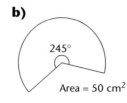
245°
Area = 50 cm²

14 Joni blows up a spherical balloon until it has a radius of 12 cm. Find the volume of air she has blown into the balloon.

15

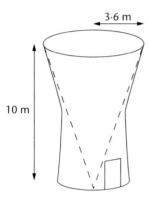
3·6 m
10 m

A concrete water tower has its internal volume in the shape of an inverted cone. The radius of the top is 3·6 m. The depth of the cone is 10 m. Calculate the volume of water which can be stored in the tower.

16 Paul measures out 250 g flour, 150 g butter, 120 g sugar, all to the nearest 10 g. Calculate the upper and lower bounds of the total mass of flour, butter and sugar.

17 Two pieces of string measure 19·7 cm and 11·4 cm respectively, to the nearest millimetre.

Calculate:
a) the upper bound of the total length of the two pieces placed end to end
b) the lower bound of the difference in the lengths of the two pieces.

18 The space for some base kitchen units is measured as 1000 mm to the nearest millimetre. Two base units are 500 mm each, to the nearest millimetre.
a) Explain why the two units will not necessarily fit into the space.
b) Calculate the upper bound of the gap remaining if two units do fit in.

19 The length of a side of a cube is given as 4·6 cm to the nearest millimetre. Calculate the upper and lower bounds of the volume of the cube, giving your answers to 3 significant figures.

20 A 100 m race was won in a time of 13·62 s, correct to the nearest hundredth of a second. Calculate, to 3 significant figures, the upper bound of the average speed
a) if the distance is 100 m to 3 significant figures
b) if the distance is 100·0 m to 4 significant figures

21 Jane walks on an exercise machine for 7·2 minutes at a speed of 130 m per minute, both measurements to 2 significant figures. Calculate the upper bound of the distance she walked.

22 A town has a population of 94 300 to the nearest 100. Its area is 156 km², to the nearest square kilometre. Calculate the lower bound of its population density, giving your answer to 3 significant figures.

8 Volumes and surface areas of similar figures

You should already know

- how to find areas and volumes
- how to use scale factors.

 This cube has a volume of $8\,\text{cm}^3$.

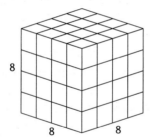 This cube has a volume of $512\,\text{cm}^3$.

The lengths of the small cube have been enlarged with scale factor 4.

The volume has been enlarged with scale factor 64.

Since there are 3 dimensions for volume, and each dimension has been enlarged with scale factor 4, the volume scale factor = 4^3.

Similarly, considering the area of the face of each cube.

For the small cube, the area is $4\,\text{cm}^2$.

For the large cube, the area is $64\,\text{cm}^2$.

The area has been enlarged with scale factor 16.

There are two dimensions for area, so the area scale factor is 4^2.

For mathematically similar shapes:

area scale factor = (length scale factor)^2

volume scale factor = (length scale factor)^3.

EXAMPLE 1

A model aircraft is made to a scale of 1:50. The area of the wing on the model is 18 cm². What is the area of the wing on the real aircraft?

Length scale factor $= 50$

Area scale factor $= 50^2$

Area of real wing $= 18 \times 50^2$

$= 45\,000\,\text{cm}^2$

$= 4.5\,\text{m}^2$.

There is an alternative solution to this problem.

With this scale, 2 cm represent 1 m.

So 4 cm² represent 1 m².

So 18 cm² represent $\dfrac{18}{4} = 4.5\,\text{m}^2$.

EXAMPLE 2

A jug holding 50 cl is 12 cm high. A similar jug holds 2 litres. What is its height?

$$50\,\text{cl} = 0.5 \text{ litres}$$

Volume scale factor $= \dfrac{2}{0.5} = 4$

Length scale factor $= \sqrt[3]{4}$

Height of larger jug $= 12\,\text{cm} \times \sqrt[3]{4}$

$= 19.0\,\text{cm}$ to 1 decimal place.

Exam tip

Remember that

$1\,\text{m} = 100\,\text{cm}$

$1\,\text{m}^2 = 10\,000\,\text{cm}^2$

$1\,\text{m}^3 = 1\,000\,000\,\text{cm}^3$

EXERCISE 8.1A

1 State the area scale factor for these length scale factors:

   a) 2   **b)** 3   **c)** 5

2 State the volume scale factor for these length scale factors:

   a) 10   **b)** 4   **c)** 5

3 State the length scale factor for the following:

   a) area scale factor of 16   **b)** volume scale factor of 216.

4 The model of a theatre set is made to a scale of 1:20. What area on the model will represent 1 m² on the real set?

5 A medicine bottle holds 125 ml. How much does a similar bottle twice as high hold?

6 A tray has an area of 160 cm². What is the area of a similar tray whose lengths are one and a half times as large?

7 To what scale is a model drawn if an area of 5 m² in real life is represented by 20 cm² on the model?

8 Two mugs are similar. One contains twice as much as the other. The smaller one is 10 cm high. What is the height of the larger one?

9 Three similar wooden boxes have heights in the ratio 3:4:5. What is the ratio of their volumes?

Exercise 8.1A cont'd

10 A model of a room is made to a scale of 1:25.

   a) What is the real height of a cupboard which is 8 cm high on the model?

   b) What is the real area of a rug which has an area of 48 cm^2 on the model?

   c) What is the volume of a waste paper basket which has a volume of 1.2 cm^3 on the model

EXERCISE 8.1B

1 State the area scale factor for these length scale factors:

   a) 3   **b)** 6   **c)** 10

2 State the volume scale factor for these length scale factors:

   a) 2   **b)** 3   **c)** 8

3 State the length scale factor for the following:

   a) area scale factor of 64

   b) volume scale factor of 1000.

4 A model of a building is made to a scale of 1:50. A room in the model is a cuboid with dimensions 7·4 cm by 9·8 cm by 6·5 cm high. Calculate the floor area of the room in:

   a) the model **b)** the actual building.

5 A glass holds 15 cl. The heights of this and a larger similar glass are in the ratio 1:1·2. Calculate the capacity of the larger glass.

6 The area of a table is 1·3 m^2. Calculate the area of a similar table twice as large.

7 The volumes of two similar jugs are in the ratio 1:4. What is the ratio of their heights?

8 9m^2 of fabric are required to cover a small sofa. What area of fabric is required to cover a similar sofa 1·2 times as long?

9 Two vases are similar. The capacity of the smaller one is 250 ml. The capacity of the larger one is 750 ml. The height of the larger one is 18 cm. Calculate the height of the smaller one.

10 The area of a rug is 2·4 times as large as the area of a similar rug. The length of the smaller rug is 1·6 m. Find the length of the larger rug.

Key idea

● For mathematically similar shapes: area scale factor = (length scale factor)^2

                        volume scale factor = (length scale factor)^3

9 Probability

Probability of event A or event B happening

Look at the grid for throwing two dice.

Suppose Louise needs a score of 8 or 11. Out of a total of 36 possible outcomes, there are 7 that give 8 or 11.

The probability of scoring 8 or 11 is therefore $\frac{7}{36}$.

But the probability of scoring 8 is $\frac{5}{36}$ and the probability of scoring 11 is $\frac{2}{36}$ and $\frac{5}{36} + \frac{2}{36} = \frac{7}{36}$.

So P(8 *or* 11) = P(8) + P(11).

If the two events are 'scoring a double' or 'scoring 8' the situation is different.

There are 10 outcomes that give a double or a score of 8 and therefore:

$$P(\text{double } or\ 8) = \frac{10}{36} \qquad P(\text{double}) = \frac{6}{36}$$

$$P(8) = \frac{5}{36} \text{ and } \frac{6}{36} + \frac{5}{36} \text{ does not equal } \frac{10}{36}.$$

So P(double *or* 8) does *not* equal P(double) + P(8).

This is because the events 'scoring a double' and 'scoring 8' are not mutually exclusive events. It is possible to do both by throwing a double 4.

The addition rule only applies to mutually exclusive events.

If events A and B are mutually exclusive then P(A or B) = P(A) + P(B).

Independent events

If two coins are tossed, the way the first one lands cannot possibly affect the way the second one lands.

Similarly, if two dice are thrown, the way the first one lands cannot possibly affect the way the second one lands.

If there are six red balls and four black balls in a bag, and one is selected and replaced before a second one is selected, the probability of getting a red ball is exactly the same on the second choice as on the first: $\frac{6}{10}$.

> When an event is unaffected by what has happened in another event, the events are said to be **independent**.

In the example of six red balls and four black ones, if the first ball is not replaced then the probability of getting a red ball on the second draw is no longer $\frac{6}{10}$ as there are fewer balls in the bag.

> When an event is affected by what has happened in another event, the events are said to be **dependent**.

Probability of event A and event B happening

Fatima tossed a coin and threw a die. Since there are 12 equally likely outcomes, and scoring a head and a 6 is one of them, it can be concluded that:

P(head and a 6) = $\frac{1}{12}$.

Now P(head) = $\frac{1}{2}$ and P(6) = $\frac{1}{6}$.

But $\frac{1}{2} \times \frac{1}{6} = \frac{1}{12}$ so P(head and a 6) = P(head) × P(6).

When tossing two coins

P(2 heads) = P(head) × P(head) = $\frac{1}{2} \times \frac{1}{2} = \frac{1}{4}$.

These results are only true because the events are independent. If they were dependent events, the second probability in the multiplication sum would be different.

> For independent events P(A *and* B) = P(A) × P(B).

Clearly it is more of a coincidence to throw two heads than one, so it is to be expected that the probability will be less. Multiplying fractions and decimals less than one gives a smaller answer, whereas adding them gives a bigger answer.

Chapter 9 *Probability*

The result for events A and B extends to more than two events. When tossing three coins, the probability of getting all three heads $= \frac{1}{8}$.

$P(\text{head}) \times P(\text{head}) \times P(\text{head}) = \frac{1}{2} \times \frac{1}{2} \times \frac{1}{2} = \frac{1}{8}$.

So, included in the multiply rule are words such as 'both' and 'all'.

Exam tip

It is very common for examination candidates to add probabilities when they should have multiplied. If you get an answer to a probability question that is more than 1, you have almost certainly added instead of multiplied.

EXAMPLE 1

The probability that the school hockey team will win their next match is 0·4. The probability that they will draw their next match is 0·3. What is the probability that they will win or draw their next match?

The events are mutually exclusive, since they cannot both win and draw their next match, so:

$P(\text{win } or \text{ draw}) = P(\text{win}) + P(\text{draw}) = 0\cdot3 + 0\cdot4 = 0\cdot7$.

EXAMPLE 2

Matt spins the fair spinner shown in the picture twice.

What is the probability that Matt scores a 4 on both his spins?

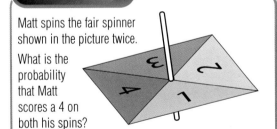

The events are independent, since the second spin cannot be affected by the first.

$P(4 \text{ } and \text{ } 4) = P(4) \times P(4) = \frac{1}{4} \times \frac{1}{4} = \frac{1}{16}$.

EXAMPLE 3

There are six red balls and four black balls in a bag. Gina selects a ball, notes its colour and replaces it. She then selects another ball. What is the probability that Gina selects:

a) two red balls

b) one of each colour?

Since Gina replaces her first ball before choosing the second the events are independent.

a) $P(2 \text{ reds}) = P(\text{red}) \times P(\text{red}) = \frac{6}{10} \times \frac{6}{10} = \frac{36}{100} = \frac{9}{25}$

(or in decimals $0\cdot6 \times 0\cdot6 = 0\cdot36$).

b) Before doing this question it is important to think about what the outcomes are.

Gina requires first ball red *and* second ball black *or* first ball black *and* second ball red.

Both the add and multiply rules are needed.

$P(\text{one of each colour}) = \left(\frac{4}{10} \times \frac{6}{10} \right) + \left(\frac{6}{10} \times \frac{4}{10} \right)$

$= \frac{24}{100} + \frac{24}{100} = \frac{48}{100} = \frac{12}{25}$

(or in decimals $(0\cdot4 \times 0\cdot6) + (0\cdot6 \times 0\cdot4) =$
$0\cdot24 + 0\cdot24 = 0\cdot48$).

Questions like part b) of Example 3, which require both rules, are clearly more difficult. Later in the chapter, you will see that these can often be more easily tackled using tree diagrams.

EXERCISE 9.1A

1 There are 4 red counters, 5 white counters and 1 blue counter in a bag. If a counter is chosen at random, find the probability that it is red or blue.

2 Eileen is choosing her next car. The probability that she chooses a Ford is 0·5, the probability that she chooses a Rover is 0·35 and the probability that she chooses a Vauxhall is 0·15. Find the probability that Eileen chooses either a Ford or a Vauxhall.

3 There are 4 aces and 4 kings in a pack of 52 playing cards. I choose a card at random from the pack. What is the probability that it is an ace or a king?

4 There are 4 red counters, 5 white counters and 1 blue counter in a bag. I choose a counter at random, note its colour and put it back in the bag. I then do this a second time. Find the probability that both my choices are red.

5 The probability that I choose chips for school dinner is 0·6. The probability that I choose pizza for school dinner is 0·4. Assuming that the events are independent, what is the probability that I choose chips on Monday and pizza on Tuesday for school dinner?

6 What is the probability that I get a multiple of 2 when I throw a single fair dice? If I throw the dice twice, what is the probability that both throws give me a multiple of 2?

7 There are 12 picture cards in a pack of 52 playing cards. Geta selects a card at random, returns it to the pack and then randomly selects another card. Find the probability that both of Geta's selections were picture cards.

8 Emma and Rebecca are choosing where to go. The probability that Emma chooses to go to the cinema is 0·7. The probability that Rebecca chooses to go to the cinema is 0·8. Assuming that their choices are independent, find the probability that they both choose to go to the cinema.

EXERCISE 9.1B

1 The probability that the school football team will win their next game is 0·65. The probability that they will draw the next game is 0·2. What is the probability that they will win or draw their next game?

2 Assuming that the results of the football team are independent, use the probabilities in Question 1 to find the probability that they will draw both of their next two games.

Exercise 9.1B cont'd

3 On Saturday morning Beverly watches TV, plays on her computer or goes to her friend's house. She says that the probability that she watches TV is 0·4, plays on her computer is 0·25 and goes to her friend's house is 0·3. Why is her statement incorrect?

4 Andy is selecting a main course and a pudding from this menu.

MENU

Main course	Pudding
Pizza (0·45)	Ice Cream (0·7)
Burger (0·3)	Fruit (0·3)
Fish Fingers (0·25)	

The numbers next to the items are the probabilities that Andy chooses these items.

a) Find the probability that Andy chooses Pizza or Burger for his main course.

b) Assuming his choices are independent, find the probability that Andy chooses Fish Fingers and Ice Cream.

5 There are 4 aces in a pack of 52 playing cards. I pick a card at random, replace it and then pick another card at random.

a) Find the probability, as a fraction in its simplest form, that the first card chosen is an ace.

b) Find the probability that both cards are aces.

6 The weather forecast says that there is a 60% chance that it will rain today and a 40% chance that it will rain tomorrow.

a) Write 60% and 40% as decimals.

b) Find the probability that it will rain on the two days.

7 There is an equal likelihood that someone is born in any month of the year. What is the probability that two people are both born in January?

8 I spin this 3-sided spinner three times. What is the probability that all my spins land on 2?

Using tree diagrams for unequal probabilities

Look again at Andy's choices on the menu in question 4 in the last exercise. These can be shown on a tree diagram with the probabilities written on the branches.

MENU

Main course	Pudding
Pizza (0·45)	Ice Cream (0·7)
Burger (0·3)	Fruit (0·3)
Fish Fingers (0·25)	

Main Course   Sweet

Pizza
— 0·7 — Ice Cream   0·45 × 0·7
— 0·3 — Fruit   0·45 × 0·3

0·45

0·3   Burger
— 0·7 — Ice Cream   0·3 × 0·7
— 0·3 — Fruit   0·3 × 0·3

0·25   Fish Fingers
— 0·7 — Ice Cream   0·25 × 0·7
— 0·3 — Fruit   0·25 × 0·3

So the probability of choosing Pizza and Fruit = 0·45 × 0·3 = 0·135

And the probability of choosing Burger and Ice Cream is 0·3 × 0·7

And so on.

As you go along the 'branches' of any route through the tree, multiply the probabilities. Now look at Example 4 in a different way

Exam tip

If you are going along the 'branches' of a tree diagram, multiply the probabilities. At the end, if you want more than one route through the tree, add the probabilities.

EXAMPLE 4

There are six red balls and four black balls in a bag. Gina selects a ball, notes its colour and replaces it. She then selects another ball. What is the probability that Gina selects:

a) two red balls

b) one of each colour?

A tree diagram can be drawn to show this information.

Notice that at each stage the probabilities add up to 1 and at the end all four probabilities add up to 1.

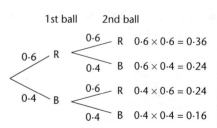

1st ball   2nd ball

0·6   R
    0·6 — R   0·6 × 0·6 = 0·36
    0·4   B   0·6 × 0·4 = 0·24

0·4   B
    0·6 — R   0·4 × 0·6 = 0·24
    0·4   B   0·4 × 0·4 = 0·16

a) Probability of red followed by red = 0·6 × 0·6 = 0·36

b) For one of each colour, Gina needs either the second route or the third route through the tree diagram.

So P(one of each colour) = (0·4 × 0·6) + (0·6 × 0·4) = 0·24 + 0·24 = 0·48

EXERCISE 9.2A

1 There are 3 red and 2 blue bricks in child's building set. The child chooses a brick at random, replaces it and then chooses another. Copy and complete the tree diagram to show the child's choices.

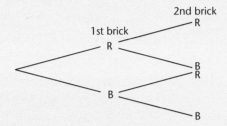

What is the probability that the child chooses:

a) 2 red bricks

b) a red brick and a blue brick in either order.

2 Andy is selecting a main course and a pudding from this menu.

> **MENU**
>
> **Main course**   **Pudding**
>
> Pizza (0·45)   Ice Cream (0·7)
> Burger (0·3)   Fruit (0·3)
> Fish Fingers (0·25)

The numbers next to the items are the probabilities that Andy chooses these items.

a) Draw a tree diagram to show Andy's choices.

b) Find the probability that Andy chooses

  (i) Burger and Ice Cream

  (ii) Fish Finger and Fruit.

3 The probability that Fatou is late for work is 0·2. Copy and complete the tree diagram for the first two days of the week.

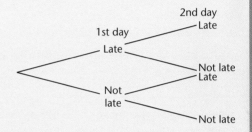

Calculate the probability that Fatou is:

a) late on both days

b) late on one of the two days.

4 There are 5 red counters, 2 blue counters and 3 yellow counters in a bag. I choose a counter at random, note its colour, replace it and then choose another.

a) Draw a tree diagram to show the results of my two choices.

b) Calculate the probability that I choose:

  (i) two red counters

  (ii) two counters of the same colour.

Exercise 9.2A cont'd

5 I toss a coin 3 times. Draw a tree diagram to show the possible outcomes. Calculate the probability that I get:

a) three heads

b) two tails and one head.

EXERCISE 9.2B

1 The probability that Mike scores a goal from a penalty kick is 0·7. Copy and complete the tree diagram for the outcomes of 2 penalty kicks.

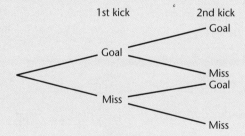

Calculate the probability that:

a) Mike scores a goal with both kicks

b) Mike scores just one goal from the two kicks.

2 Qasim arrives at school on time 40 out of 50 days in any term.

a) Write down, as a fraction in its simplest form, the probability that Qasim is late for school.

b) Draw a tree diagram to show whether Qasim is on time for school on 2 days.

c) Find the probability that Qasim will arrive on time at school:

(i) on both days

(ii) on one of the days.

3 Extend the tree diagram you drew for question 2 to show whether Qasim is late for school on 3 days. Find the probability that Qasim is on time on:

a) all 3 days

b) one of the 3 days.

4 There are 12 red counters, 6 white counters and 2 blue counters in a box. I choose one counter at random, note its colour, replace it and then choose another.

a) Draw a tree diagram to show the results of my choices.

b) Calculate the probability that I choose:

(i) two red counters

(ii) two counters of the same colour

(iii) two counters of different colours.

Exercise 9.2B cont'd

5 Sarah drew this tree diagram for choosing a person with a certain hair colour from her class.

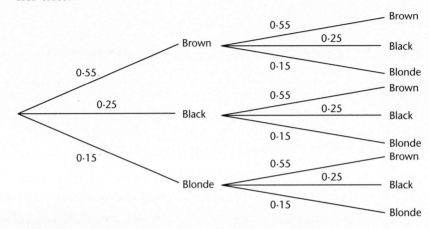

Explain why Sarah has made a mistake.

Conditional probability

Earlier in this chapter you learnt this rule for independent events A and B:

P(A and B) = P(A) × P(B).

For instance, if you toss a coin and a die, the probability of a head is $\frac{1}{2}$ and the probability of a six is $\frac{1}{6}$.

The events are independent since what happens to the coin cannot possibly affect what happens to the die.

So the probability of getting a head and a six = $\frac{1}{2} \times \frac{1}{6} = \frac{1}{12}$.

> There are many events, however, where the outcome of the second event is affected by the outcome of the first event. In this situation the probability of the second event depends on what has happened in the first event.
>
> These events are not independent and are thus called **dependent events**.
>
> In this situation the probability of the second event is called a **conditional probability** since it is conditional on whether the first event has happened or not.

In some situations you can work out the conditional probability yourself. In others you will be told what it is. These situations are illustrated in the following examples.

EXAMPLE 5

There are 4 red balls and 6 black balls in a bag. If the first one selected is black and is not replaced, what is the probability that the second one is also black?

There are now 4 red and 5 black balls left in the bag so the probability is $\frac{5}{9}$.

EXAMPLE 6

There are 7 blue balls and 3 red balls in a bag. A ball is selected at random and not replaced. A second ball is then selected.

Find the probability that **a)** 2 blue balls are chosen   **b)** 2 balls of the same colour are chosen.

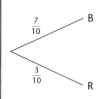

1st Ball

A tree diagram is still one of the best ways to organise the work and the first step is the same as in the replacement situation.

For the second ball, however, the situation is different. If the first ball was blue there are now 6 blue balls and 3 red balls left in the bag. The probabilities are therefore $\frac{6}{9}$ and $\frac{3}{9}$.

If the first ball was red there are now 7 blue and 2 red balls left in the bag. The probabilities are therefore $\frac{7}{9}$ and $\frac{2}{9}$.

The tree diagram now looks like this

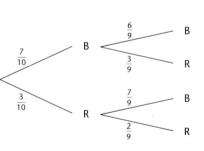

a) The probability that the 2 balls are both blue
$$= \frac{7}{10} \times \frac{6}{9} = \frac{42}{90} = \frac{7}{15}$$

b) The probability that both balls are the same colour (i.e. both blue *or* both red)
$$= \frac{7}{10} \times \frac{6}{9} + \frac{3}{10} \times \frac{2}{9} = \frac{42}{90} + \frac{6}{90} = \frac{48}{90} = \frac{8}{15}.$$

Notice that the multiplication part of the rule still applies. It is the probabilities that are different from the independent case.

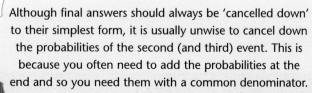

Exam tip

Although final answers should always be 'cancelled down' to their simplest form, it is usually unwise to cancel down the probabilities of the second (and third) event. This is because you often need to add the probabilities at the end and so you need them with a common denominator.

EXAMPLE 7

I have sandwiches for lunch on 70% of school days, otherwise I have a school meal. If I have sandwiches, the probability that I buy a drink from the canteen is 0·2. If I have a school meal the probability I buy a drink is 0·9.

Find the probability that I buy a drink.

Here the probabilities are given to us. The tree diagram for this situation is this.

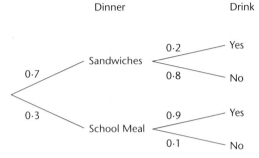

The probability of buying a drink
= 0·7 × 0·2 + 0·3 × 0·9 = 0·14 + 0·27 = 0·41.

Note that once the probabilities have been put onto the tree diagram, the methods are the same.

Exam tip

Some questions may instruct you to draw a tree diagram, others may not. Whilst it is not essential to draw a tree diagram for all questions it is a very powerful method for tackling probability problems. Even if you are not told to do so, you are well advised to draw one.

In problems involving selections from bags etc., always assume that the probabilities are **dependent** unless you are specifically told that there is replacement or that the events are independent.

EXERCISE 9.3A

1 There are 5 blue balls in a bag and 4 white ones. If a white one is selected first, what is the probability that the second ball is blue?

2 500 tickets are sold in a raffle. I buy 2 tickets. I do not win the first prize. What is the probability that I win the second prize?

3 There are 8 balls in a bag and 2 white ones. 2 balls are selected at random without replacement.

   a) Draw a tree diagram to show the probabilities of the possible outcomes.

   b) Find the probability of selecting 2 white balls.

   c) Find the probability of selecting 2 balls of different colour.

Exercise 9.3A cont'd

4 1000 tickets are sold in a raffle. I buy 2 tickets. What is the probability that I win both of the first two prizes?

5 There are 7 yellow balls and 4 red balls in a bag. 2 balls are selected at random without replacement. Find the probability of selecting at least 1 red ball.

6 There are 5 red balls, 3 yellow balls and 2 green balls in a bag. 2 balls are selected at random without replacement.

   a) Find the probability of selecting 2 red balls.

   b) Find the probability of selecting 2 balls of the same colour.

7 On average, Gary takes sandwiches to school for lunch on 2 days a week, otherwise he has a school meal. If he takes sandwiches the probability that he has time to play football is 0·8. If he has a school meal the probability that he has time to play football is 0.3

   a) Draw a tree diagram to show the probabilities of the possible outcomes.

   b) Find the probability that on any given day Gary has time to play football.

8 The probability that the school will win the first hockey match of the season is 0·6. If they win the first match of the season the probability that they win the second is 0·7, otherwise it is 0·4.

   a) Copy and complete the tree diagram

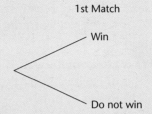

1st Match

Win

Do not win

   b) Find the probability that they win 1 of the first 2 matches only.

   c) Find the probability that they win at least 1 of the first 2 matches.

9 Salima walks to school, cycles or goes by bus. The probability that she walks is 0·5. The probability that she cycles is 0·2. If she walks, the probability that she is late is 0·2. If she cycles, it is 0·1 and if she goes by bus, it is 0·4.

   a) Draw a tree diagram to show the probabilities of the possible outcomes.

   b) Find the probability that, on any given day, she is on time for school.

10 If Ryan is fit to play, the probability that United will win their next match is 0·9. If he is not fit to play, the probability is 0·8. The physiotherapist says there is a 60% chance he will be fit.

   What is the probability that United will win their next match?

EXERCISE 9.3B

1 Soraya and Robert are selecting cards in turn from an ordinary pack of 52 playing cards. Soraya selects a heart and does not replace it. What is the probability that Robert also selects a heart?

2 There are 80 tulip bulbs and 120 daffodil bulbs in a tub. Alan chose 2 bulbs and there was 1 of each. Charlie then chose a bulb. What is the probability that Charlie's bulb was a tulip?

3 There are 3 red pens and 5 blue ones in a pencil case. Jenny chose 2 pens at random.

   a) Draw a tree diagram to show the probabilities of Jenny's possible choices.

   b) Find the probability that Jenny chose at least 1 blue pen.

4 There are 5 grey socks, 3 black socks and 4 navy socks in Lisa's drawer. She selects 2 socks at random.

   What is the probability that she selects a pair of the same colour socks?

5 Sanjay selects 3 cards without replacement from a normal pack of 52 playing cards.

   What is the probability that he selects 3 aces?

6 There are 5 men and 4 women on a committee. 2 are selected at random to represent the committee on a working party. What is the probability that the two selected are:

   a) both women   **b)** both men   **c)** one woman and one man?

7 In a game I toss a coin and spin one of these fair spinners

   If I toss a head I spin the five-sided spinner.

   If I toss a tail I spin the six-sided spinner.

   I need a 5 to win the game.

   What is the probability that I win the game?

   (**Hint:** If you use a tree diagram, take the events for the spinner as '5' and 'not a 5'.)

Exercise 9.3B cont'd

8 If it is fine when he gets up there is a probability of 0·7 that Richard will cycle to school. If it is raining when he gets up there is a probability of 0·05 that he cycles. The weatherman estimates that there is a 20% chance that it will rain tomorrow morning.

Using the weatherman's estimate, find the probability that Richard will cycle to school tomorrow.

9 Paul chooses 2 cards without replacement from an ordinary pack of 52 playing cards.

Find the probability that he chooses:

a) 2 hearts **b)** 2 kings **c)** a king and a queen.

10 There are 7 blue balls and 3 red ones in a bag. Rosemary selects 3 balls at random without replacement.

a) Draw a tree diagram to show the probabilities of the possible outcomes.

b) Find the probability that Rosemary chooses at least 1 red ball.

c) Find the probability that she chooses 2 blue balls and 1 red ball.

Key ideas

- If events A and B are mutually exclusive, then P(A or B) = P(A) + P(B).
- For independent events, P(A and B) = P(A) × P(B).
- The multiply rule should also be used for words like 'both' and 'all'.
- If the outcome of the second event is affected by the outcome of the first event, the probability of the second event will vary according to what happens in the first event.
- These events are called **dependent events**.
- In this situation the probability of the second event is called a **conditional probability** since it is conditional on whether the first event has happened or not.
- Once the conditional probabilities have been established, they can be placed on a tree diagram and the multiplication and addition rules still apply.

10 Problems in 3D

You should already know

- how to apply Pythagoras' theorem
- how to use trigonometry in right-angled triangles
- the convention for labelling sides and angles in a triangle.

EXAMPLE 1

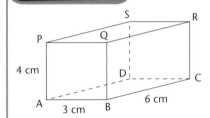

The rectangular box measures 4 cm by 3 cm by 6 cm. Calculate:

a) AC

b) AR

c) angle RBC

d) angle ARC

Exam tip

Identify the triangle required and draw it out separately. Label the corners and mark on any lengths and angles known.

a)

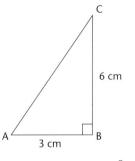

$AC^2 = AB^2 + RC^2$

$AC^2 = 3^2 + 6^2$

$AC^2 = 45$

$AC = 6.71 \, cm$

Exam tip

Don't use 6.71^2 – this has been rounded. You need AC^2 and this is 45.

b)

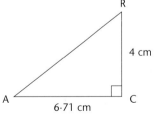

$AR^2 = AC^2 + RC^2$

$AR^2 = 45 + 4^2$

$AR^2 = 61$

$AR = 7.81 \, cm$

Exam tip

When there is a choice of trig. formulas to use, use the one which contains as many given values as possible. You may have calculated a value incorrectly.

Example 1 cont'd

c)

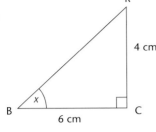

$$\tan x = \frac{4}{6}$$

$$x = \tan^{-1} \frac{4}{6}$$

$$x = 33 \cdot 7°$$

d)

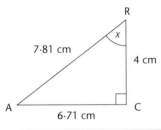

$$\cos x = \frac{4}{7 \cdot 81...}$$

$$x = \cos^{-1} \frac{4}{7 \cdot 81}$$

$$x = 59 \cdot 2°$$

Exam tip

You can't avoid using a calculated value here. Using $7 \cdot 81$ will give an answer correct to 3 significant figures but it may be worth using a more accurate value, $\sqrt{61}$ for instance. Round at the end.

EXAMPLE 2

A tree, TC, is 20 m North of point A. The angle of elevation of the top of the tree, T, from A, is 35°. A point, B, is 30 m East of point A. A, B and C are on horizontal ground. Calculate:

a) the height of the tree, TC

b) the length, BC

c) the angle of elevation of T from C.

Exam tip

First, a diagram of the situation is needed.

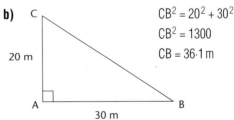

a)

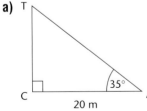

$$\tan 35 = \frac{TC}{20}$$

$$TC = 20 \tan 35$$

$$TC = 14 \cdot 0 \, m$$

b)

$$CB^2 = 20^2 + 30^2$$

$$CB^2 = 1300$$

$$CB = 36 \cdot 1 \, m$$

Example 2 cont'd

c)

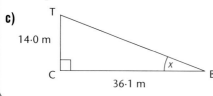

$$\tan x = \frac{14 \cdot 0 \ldots}{36 \cdot 1 \ldots}$$

$$x = \tan^{-1}\left(\frac{14 \cdot 0 \ldots}{36 \cdot 1 \ldots}\right)$$

$$x = 21 \cdot 2°$$

EXAMPLE 3

A mast, MG, is 50 m high. It is supported by two ropes, AM and BM, as shown.

ABG is horizontal.

Other measurements are shown on the diagram.

Is the mast vertical?

Does $\sin 38 \cdot 7° = \frac{50}{80}$? $\arcsin\left(\frac{50}{80}\right) = 38 \cdot 68 \ldots$

Yes, angle MGA is a right angle.

Is angle MGB a right angle?

$$50^2 + 50^2 = 5000$$

$$\sqrt{5000} = 70 \cdot 7 \ldots$$

No, MB is shorter than this.

The mast leans towards B (angle MGB < 90°).

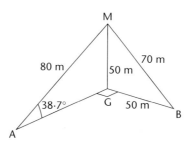

It is sometimes helpful to use 3D coordinates in these problems.

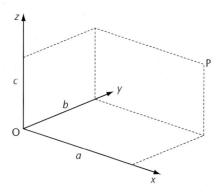

The coordinates of P are (a, b, c).

EXAMPLE 4

In this cuboid, OA = 5, AB = 3 and EB = 1.

Write down the coordinates of the vertices.

O(0, 0, 0), A(5, 0, 0), B(5, 3, 0), C(0, 3, 0),
D(5, 0, 1), E(5, 3, 1), F(0, 3, 1), G(0, 0, 1).

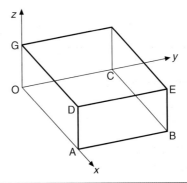

EXERCISE 10.1A

1 ABCDEFGH is a rectangular box with dimensions as shown.

   a) Calculate:

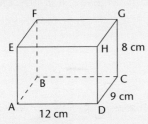

   (i) angle GDC

   (ii) EG

   (iii) EC

   (iv) angle GEC

   b) Take A as origin, AD as *x*-axis, AB as *y*-axis and AE as *z*-axis. Write down the
   coordinates of the vertices.

2 A wedge has a rectangular base, ABCD, on horizontal ground. The rectangular face,
   BCEF, is vertical. Calculate:

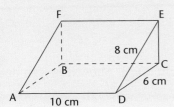

   a) FC

   b) angle DFC

   c) ED

   d) angle EDB

Exercise 10.1A cont'd

3 Three points, A, B and C, are on horizontal ground with B due West of C and A due South of C. A chimney, CT, at C, is 50 m high. The angle of elevation of the top, T, of the chimney is 26° from A and 38° from B. Calculate:

a) how far A and B are from C

b) the distance between A and B

c) the bearing of B from A.

4 In the cuboid, PQ = 7·5 cm, QR = 4 cm and PX = 12 cm. Calculate:

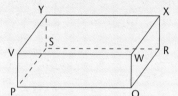

a) XR

b) angle PXR.

5 The figure shows a square-base pyramid with V vertically above the centre, X, of the square ABCD. Given that AB = 8 cm and AV = 14 cm, calculate:

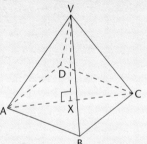

a) angle CAB

b) angle VCB. (**hint:** triangle VCB is isosceles)

c) AC

d) AX

e) VX

f) angle VAX.

6 PQRSTU is a triangular prism

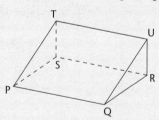

PQ = SR = TU = 5·7 cm.

PT = QU = 4·3 cm, PR = 6·9 cm.

Calculate:

a) PU

b) QR.

UR = 2·1 cm.

c) Is angle URQ a right angle? Show how you decide.

EXERCISE 10.1B

1 ABCDPQ is a triangular prism with ABPQ horizontal and ADP vertical.

a) Calculate:

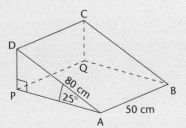

    (i) DP

    (ii) AC

    (iii) angle CAQ

    (iv) the volume of the prism.

b) Take A as origin, AB as x-axis, AD as y-axis and AR as z-axis. Write down the coordinates of the vertices.

2 A pyramid has a rectangular base, ABCD, with AB = 15 cm and BC = 8 cm. The vertex, V, of the pyramid is directly above the centre, X, of ABCD with VX = 10 cm.

a) Calculate:

    (i) AC

    (ii) AV

    (iii) angle AVB.

b) Given that M is the mid-point of AB, calculate:

    (i) VM

    (ii) angle VMX.

3 ABCDEFGH is a cuboid. AB = 7 cm, AC = 8·6 cm and angle GBC = 41°. Calculate:

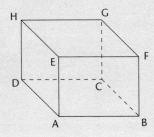

    a) BC

    b) angle GAC.

Exercise 10.1B cont'd

4 Triangle ABC is horizontal. X is vertically above A. AC = AX = 15 cm.
Angle ACB = 27° and angle BAC = 90°. Calculate:

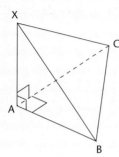

a) XC

b) BC

c) BX.

5 The figure shows a garden shed with the floor horizontal and the walls vertical.
Calculate:

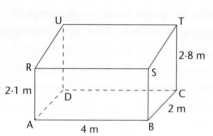

a) AT

b) ST

c) the angle that the roof of the shed makes
with the horizontal

d) RT.

6 A field is a quadrilateral with opposite sides equal.

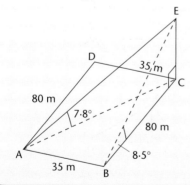

AD = BC = 80 m.

DC = AB = 35 m.

A vertical post, CE, is at one corner of the field.
The angles of elevation of the top of the post
are 7·8° from A and 8·5° from B.

Is the field a rectangle? Show how you decide.

The angle between a line and a plane

In the previous exercises of this chapter, you have been finding the angle between a line and a plane when it has been specified by required letters. This next section deals with finding the angle when only the line and the plane are given.

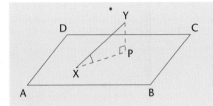

The end, X, of the line XY is on the plane ABCD. The angle between the line and the plane is the angle YXP, where P is the point on the plane vertically below Y.

EXAMPLE 5

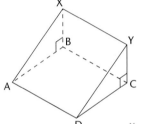

For this triangular prism, sketch the triangle and label the angle between the line and the plane given.

a) DY and ABCD

b) AY and ABCD

c) AY and BCXY

d) BY and ABX

a) Angle YDC

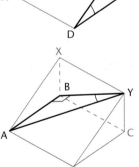

b) Angle YAC

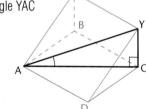

c) Angle AYB

d) Angle YBX

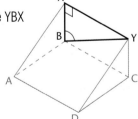

77

EXAMPLE 6

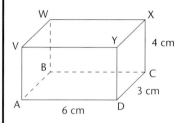

ABCDVWXY is a cuboid. Calculate the angle between the following lines and planes.

a) DW and ABCD

b) DW and ABVW

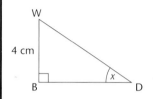

a) Angle BDW is the required angle.

> ### First find length BD.

$BD^2 = 6^2 + 3^2$

$BD^2 = 45$

$BD = 6·71 \, cm$

$\tan x = \dfrac{4}{6·71 \ldots}$

$x = \tan^{-1} \dfrac{4}{6·71 \ldots}$

$x = 30·8$

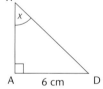

b) Angle DWA is the required angle.

> ### First find length WA.

$WA^2 = 3^2 + 4^2$

$WA^2 = 25$

$WA = 5 \, cm$

$\tan x = \dfrac{6}{5}$

$x = \tan^{-1} \dfrac{6}{5}$

$x = 50·2°$

EXERCISE 10.2A

1 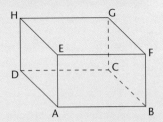 For this cuboid, sketch the triangle and label the angle between the following lines and planes.

    a) EB and ABCD

    b) EB and ADHE

    c) AG and ABCD

    d) AG and CDHG

2 Given that, for the cuboid in question 1, AB = 8 cm, BC = 6 cm and GC = 5 cm, calculate the angles between the lines and the planes listed above.

3 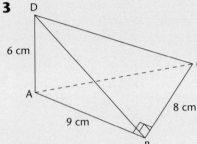 The diagram is of a tetrahedron.

ABC is a right-angled triangle on a horizontal plane. D is vertically above A and angle DBC = 90°.

Calculate:

    a) the angle between BD and ABC

    b) the angle between DC and ABD.

EXERCISE 10.2B

1 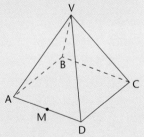 In the pyramid, ABCD is a rectangle and V is vertically above the centre of the rectangle. M is the mid-point of AD. Sketch the triangle and label the angle between the following lines and planes:

    a) VC and ABCD

    b) VM and ABCD.

2 Given that, for the pyramid in question 2, AD = 8 cm, DC = 6 cm and that VA = VB = VC = VD = 12 cm, calculate the angles between the lines and the planes listed above.

Chapter 10 *Problems in 3D*

Exercise 10.2B cont'd

3

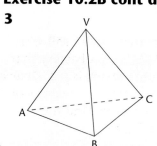

The diagram is of a tetrahedron. AB = AC = BC = 4 m and VA = VB = VC = 6 m. Calculate the angle that line VA makes with plane ABC.

Hint: V lies directly above a point, X, which is $\frac{2}{3}$ of the way from A, along the bisector of angle CAB.

Key ideas

- Identify the triangle containing the unknown side or angle. Sketch it out separately.
- Use Pythagoras' theorem and trigonometry to find the answer.

Revision exercise

C1

1 **a)** Two triangles are similar. One has a base 6 cm long and an area of 20 cm². The other has a base 18 cm long. What is its area?

b) A cone has a height of 12 cm, a surface area of 380 cm² and a volume of 480 cm³. A similar cone has a volume of 60 cm³. What are its height and surface area?

c) Three similar cylindrical cans have diameters 5 cm, 8 cm and 10 cm. The smallest can holds 150 ml. What is the capacity of the other two cans?

2 Two bottles are similar. Their heights are in the ratio 1:1·5. The larger one holds 2700 ml. What does the smaller one hold?

3 The areas of two similar pieces of paper are in the ratio 1:8. The larger piece of paper is 21·0 cm wide. What is the width of the smaller piece?

4 The probability that Zoe is late for school is 0·2. The probability that she is on time for school is 0·65. What is the probability that Zoe is either on time or late for school?

5 Assuming that the events in question 4 are independent, find the probability that Zoe is late for school on Monday and on time for school on Tuesday.

6 A fairground game offers a bottle of wine if you can throw a 6 with fair dice on two successive throws. What is the probability of this happening?

7 The probability that United will win any match is 0·7. The probability that they will draw any match is 0·2. Draw a tree diagram to show the outcomes of their next two matches. Find the probability that United:

a) lose both matches

b) do not lose both matches

c) win one of the matches and draw the other.

8 The probability that it is sunny on any day in January is 0·3. Find the probability that for two days in January:

a) both are sunny

b) one of the days is sunny.

9 The probability that the school netball team will win any match is 0·4. The probability that they draw any match is 0·1. Draw a tree diagram to show the outcomes of their next two matches. Find the probability that the team:

a) loses both matches

b) does not lose both matches

c) wins one of the two matches and draws the other.

10 The probability that it rains on 15 July is 0·1. The probability that it rains on 16 July is also 0·1.

Find the probability that it:

a) rains on both days

b) rains on one of the two days.

11

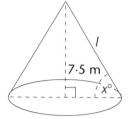

7·5 m

l

x°

A street light suspended 7·5 m above the ground illuminates a circle with circumference 30·8 m. Calculate:

a) the angle marked $x°$

b) the length l.

12

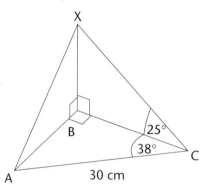

X

B

25°

38°

C

A

30 cm

In the diagram, triangle ABC is horizontal with angle ABC = 90°. XB is vertical. Find the length of XB.

13

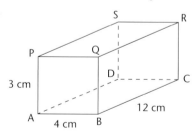

S

R

P

Q

D

C

3 cm

12 cm

A

4 cm

B

ABCDPQRS is a cuboid.

a) Calculate the area of the rectangle PQCD.

b) Which line, BS or AS, is more steeply inclined to the base ABCD?

c) Calculate the volume of the pyramid ABCDR. (**Hint:** volume of pyramid $= \frac{1}{3}$ area of base × height.)

14

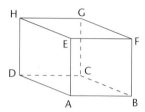

H

G

E

F

D

C

A

B

ABCDEFGH is a cube. Specify the angle between:

a) the line BH and the plane DCGH

b) the line AG and the plane BCGF.

15

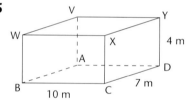

V

Y

W

X

4 m

A

D

B

10 m

C

7 m

A classroom measures 10 m by 7 m by 4 m. Calculate:

a) the angle between BY and ABWV

b) the angle between AC and ADVY.

16

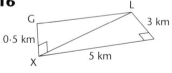

L

G

3 km

0·5 km

X

5 km

W

N

S

E

A glider, G, is 3 km east and 5 km south of its landing strip, L. It is at an altitude of $\frac{1}{2}$ km above a point X on the ground. Calculate:

a) the bearing of the point X from L

b) the distance it has to fly to reach the landing strip

c) the glider's inclination as measured from L.

11 Histograms

You should already know

- how to calculate the mean of a set of data
- how to calculate the mean for grouped data
- the meaning of the symbols ≤ and <.

Histograms

Histograms and bar charts are closely related.

- In a histogram the frequency of the data is shown by the area of each bar. (In a bar chart the frequency is shown by the height of each bar.)
- The data is continuous. (Bar charts can be for discrete data.)
- Histograms have bars, or columns, whose width is in proportion to the size of the group of data each bar represents – the class width – so the bars may have different widths. (In a bar chart the widths of each bar are usually the same.)
- The vertical scale is 'frequency density'.

Frequency density = frequency ÷ class width. (In a bar chart the vertical scale is the actual frequency.)

EXAMPLE 1

An airline investigated the ages of passengers flying between London and Johannesburg.

The table shows the findings.

Age (A years)	Frequency
$0 \leq A < 20$	28
$20 \leq A < 30$	36
$30 \leq A < 40$	48
$40 \leq A < 50$	20
$50 \leq A < 70$	30
$70 \leq A < 100$	15

Example 1 cont'd

To draw a histogram you must first calculate the frequency density.

Age (A years)	Class width	Frequency (f)	Frequency density
$0 \leq A < 20$	20	28	$28 \div 20 = 1.4$
$20 \leq A < 30$	10	36	$36 \div 10 = 3.6$
$30 \leq A < 40$	10	48	$48 \div 10 = 4.8$
$40 \leq A < 50$	10	20	$20 \div 10 = 2$
$50 \leq A < 70$	20	30	$30 \div 20 = 1.5$
$70 \leq A < 100$	30	15	$15 \div 30 = 0.5$

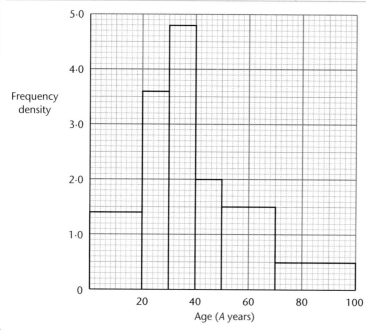

Chapter 11 *Histograms*

EXERCISE 11.1A

1 200 commuters were surveyed to find the distances they travelled to work. The table shows the results. Draw a histogram to show this information

Distance (d km)	Frequency (f)
$0 \leq d < 5$	3
$5 \leq d < 10$	9
$10 \leq d < 15$	34
$15 \leq d < 20$	49
$20 \leq d < 30$	17
$30 \leq d < 40$	41
$40 \leq d < 60$	27
$60 \leq d < 100$	20

2 The table shows the sizes of 48 marrows grown on an allotment.

Show this information on a histogram.

Length (L cm)	Frequency (f)
$0 \leq L < 20$	4
$20 \leq L < 40$	6
$40 \leq L < 50$	13
$50 \leq L < 60$	11
$60 \leq L < 90$	14

3 The table shows the results of a survey to find the areas, to the nearest hectare, of 160 farms.

Draw a histogram to show this information

Area (A hectares)	Frequency (f)
$1 \leq A < 4$	29
$4 \leq A < 8$	18
$8 \leq A < 12$	34
$12 \leq A < 16$	26
$16 \leq A < 24$	28
$24 \leq A < 30$	11
$30 \leq A < 34$	8
$34 \leq A < 40$	6

EXERCISE 11.1B

1 The age of each person on a holiday coach tour is recorded. The table shows the results.

Draw a histogram to show this information.

Age (A years)	Frequency (f)
$0 \leq A < 10$	0
$10 \leq A < 20$	2
$20 \leq A < 30$	3
$30 \leq A < 45$	8
$45 \leq A < 50$	5
$50 \leq A < 70$	18
$70 \leq A < 100$	12

2 A clothing manufacturer needs to know how long to make the sleeves of sweatshirts. 100 teenagers had their arm lengths measured. The results are shown in the table.

Draw a histogram to show this information.

Arm length (L cm)	Frequency (f)
$40 \leq L < 45$	4
$45 \leq L < 50$	22
$50 \leq L < 55$	48
$55 \leq L < 60$	14
$60 \leq L < 70$	10
$70 \leq L < 80$	2

3 An insurance company records the ages of people who were insured for holiday accidents etc. during a two-week period in August.

Draw a histogram to show this information.

Age (A years)	Frequency (f)
$0 \leq A < 5$	20
$5 \leq A < 10$	54
$10 \leq A < 20$	106
$20 \leq A < 30$	223
$30 \leq A < 40$	180
$40 \leq A < 60$	252
$60 \leq A < 90$	54

If the information is already given in a histogram it is possible to find out the frequencies and also estimate the mean.

EXAMPLE 2

This example shows how:

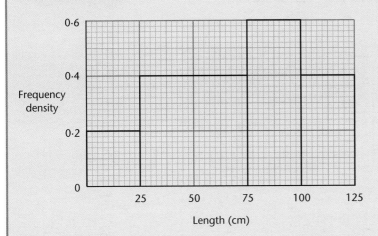

Length (cm)

The frequency density = frequency ÷ column width.

The frequencies therefore are: 0·2 × 25 = 5

0·4 × 50 = 20

0·6 × 25 = 15

0·4 × 24 = 10

The total frequency   = 50.

The mean can be estimated using the mid-point of each
class width:

$\Sigma fx = 5 \times 12·5 + 20 \times 50 + 15 \times 87·5 + 10 \times 112·5 = 3500.$

Therefore $\bar{x} = \dfrac{3500}{50} = 70.$

Chapter 11 *Histograms*

EXERCISE 11.2A

1 This histogram shows the ages of people who live in a small village.

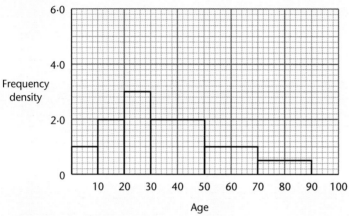

a) How many people live in the village?

b) Estimate their mean age.

2 This histogram shows the distribution of the weights of all the people living in a street.

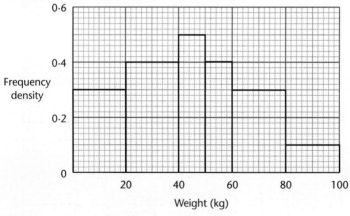

a) How many people live in the street?

b) What was the mean weight?

1 A survey of pupils in a school was made in order to find the times it took them to travel from home to school each morning. The survey was made on one particular day. The results are shown in the histogram.

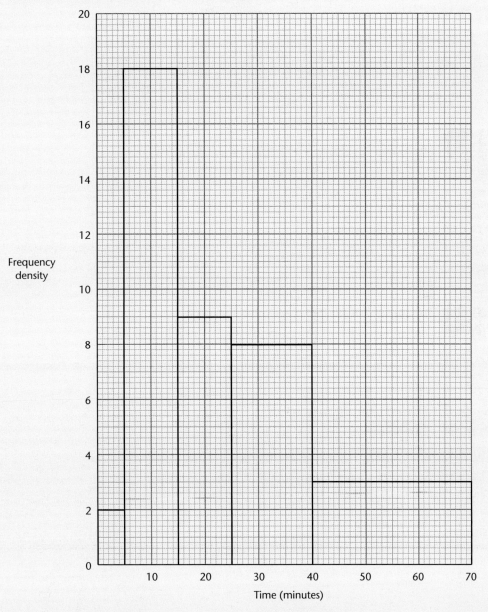

Exercise 11.2B cont'd

    a) Find the number of pupils surveyed.

    b) Calculate an estimate of the mean time it took to travel from home to school.

2 This histogram shows the results of a survey into the distance travelled by people to work each day.

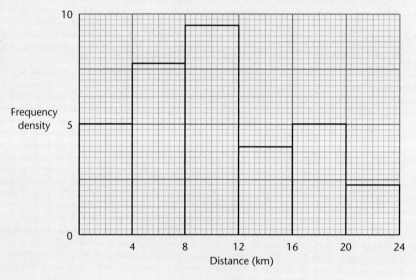

a) Find the number of people who were surveyed.

b) Calculate an estimate of the mean distance travelled to work.

Key idea

- Histograms represent frequency by the area of the columns.

12 Circle properties

You should already know

- that the exterior angle of a triangle equals the sum of the interior opposite angles
- that the sum of the angles in a triangle equals 180°
- that the sum of the angles on a straight line equals 180°
- that the opposite angles of a cyclic quadrilateral add to 180°
- about the properties of congruent triangles
- that the angle in a semi-circle equals 90°.

Angles at the centre of a circle

AB is a diameter of the circle whose centre is O. C is a point on the circumference.

Chord AC subtends (forms) angle AOC at the centre of the circle and angle ABC at the circumference.

Here is a simple proof that angle AOC = twice angle OBC.

If we draw CA and CB then, for triangle OCB

angle AOC = angle OCB + angle OBC (exterior angle of triangle).

But OC = OB.   (radii)

Therefore triangle OCB is isosceles.

Therefore angle OCB = angle OBC.

Therefore angle AOC = 2 × angle OBC.

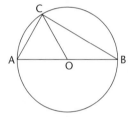

i.e. the angle at the centre = 2 × the angle at the circumference.

Exam tip

In a proof you should write out each step of your thinking and reasoning together with the reason or justification for each statement you write down.

ACTIVITY 1

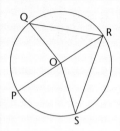

Try to prove that angle QOS = 2 × angle QRS.

(The solution is given below – try to write out your own proof and then use this to *check* your work.)

Solution

Angle POQ = angle OQR + angle ORQ (exterior angle of triangle OCR)

But OQ = QR (radii)

Therefore angle OQR = angle ORQ (opposite angles of isosceles triangle)

Therefore angle POQ = 2 × angle ORQ.

Similarly angle POS = 2 × angle ORS.

Therefore angle POQ + angle POS = 2 × angle ORQ + 2 × angle ORS

i.e. angle QOS = 2 × angle QRS.

What has just been proved can be expressed in words as:

the angle at the centre of a circle = twice the angle at the circumference subtended by the same arc (or the same chord).

(In this proof the arc is arc QPS.)

EXAMPLE

In this diagram O is the centre of the circle. Calculate the value of angle *a*.

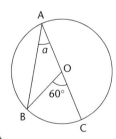

Angle *a* = 30° (angle at the centre = 2 × angle at the circumference)

In each of the following diagrams O is the centre of the circle. Find the size of each of the lettered angles. Try to write down each step with the reasons for your deductions.

1

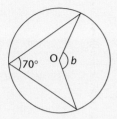

2

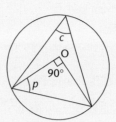

3

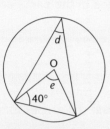

4

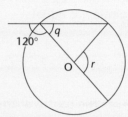

In each of the following diagrams O is the centre of the circle. Find the size of each of the lettered angles. Try to write down each step with the reasons for your deductions.

1

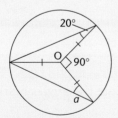

2

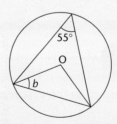

3

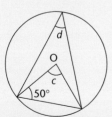

4

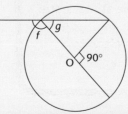

Chapter 12 *Circle properties*

ACTIVITY 2

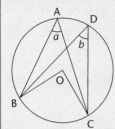

Prove that angle a = angle b.

(The solution is given below – try to write out your own proof and then use this to *check* your work.)

Solution:

angle $a = \frac{1}{2} \times$ angle BOC   (angle at centre = $\frac{1}{2}$ angle at circumference)

angle $b = \frac{1}{2} \times$ angle BOC   (angle at centre = $\frac{1}{2}$ angle at circumference)

therefore angle a = angle b.

This well-known proof is often written as: **the angles in the same segment of a circle are equal.**

EXERCISE 12.2A

In the following questions O is the centre of the circle. Calculate the angles marked with letters.

1 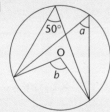   **2** 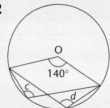   **3** 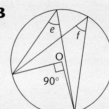   **4**

EXERCISE 12.2B

In the following questions O is the centre of the circle. Calculate the angles marked with letters.

1 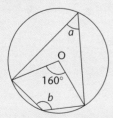   **2** 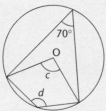   **3**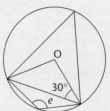

ACTIVITY 3

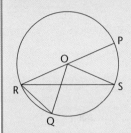

Prove angle QRS = $\frac{1}{2}$ × angle QOS.

(The solution is given below – try to write out your own proof and then use this to *check* your work.)

Solution:

Let angle ORQ = x and angle ORS = y.

Triangle ORQ is isosceles (OR and OQ are radii)

therefore angle ROQ = $180° - 2x$ (angle sum of triangle).

Triangle ORS is isosceles (OR and OS are radii)

therefore angle y = angle OSR

therefore angle ROS = $180° - 2y$ (angle sum of triangle)

angle QOS = angle ROS – angle ROQ

= $(180° - 2y) - (180° - 2x)$

= $2x - 2y$

= $2(x - y)$

but angle QRS = angle ORQ – angle ORS

= $x - y$

therefore angle QOS = 2 × angle QRS.

Sometimes the angle at the circumference is in the same semicircle as the arc.

EXERCISE 12.3A

Calculate the sizes of the angles marked with letters. O is the centre of each circle. Give the reasons for each step of your working.

1

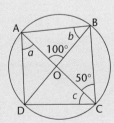

2

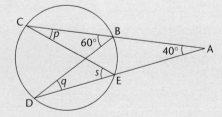

Exercise 12.3A cont'd

3

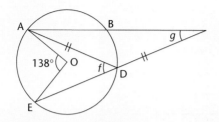

4

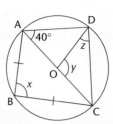

EXERCISE 12.3B

Calculate the sizes of the angles marked with letters. O is the centre of each circle.
Give the reasons for each step of your working.

1

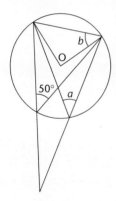

2

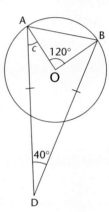

3

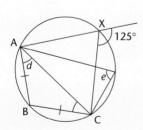

4

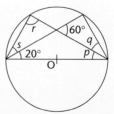

Tangents

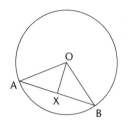

AB is a chord to the circle centre O.

X is the mid-point of AB.

Triangle AOB is isosceles.

OX is a line of symmetry for this triangle.

Therefore angle OXA = angle OXB = 90°.

What has just been shown is:

> The straight line which joins the centre of a circle to the mid-point of a chord is at right angles to the chord.

If the line OX is extended to become a radius, as in the diagram below, the chord AB will become the tangent at X. OX will be perpendicular to the tangent drawn to the circle at the point of contact, X.

This fact is proved below.

Work through the proof and make sure you can follow it.

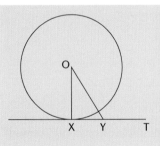

TX is the tangent to the circle centre O. OX is the radius at the point of contact. If angle OXT is not 90° it must be possible to draw a line at right angles to TX, the line OY, and so angle OYX = 90°. If angle OYX = 90° then OX is the hypotenuse of triangle OYX and OX > OY. What follows from this statement is that Y must be inside the circle because OX is a radius.

Therefore line TYX must cut the circle, i.e. line YX would be part of a chord.

This is impossible because TX is defined as a tangent,

therefore it is impossible for angle OXT not to be 90°

therefore angle OXT is 90°.

This is an example of a proof by contradiction.

EXERCISE 12.4A

In the following questions calculate the angles marked with letters. O is the centre of each circle.

X and Y are the points of contact of the tangents to each circle.

1

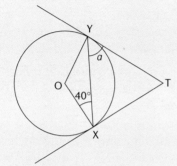

2

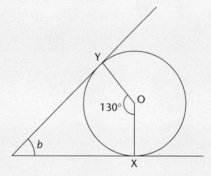

3

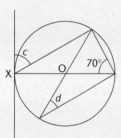

4

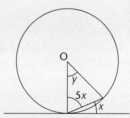

EXERCISE 12.4B

In the following questions calculate the angles marked with letters. O is the centre of each circle.

X and Y are the points of contact of the tangents to each circle.

1

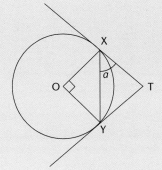

2

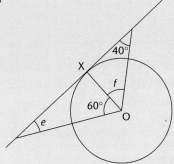

3

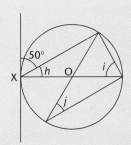

4

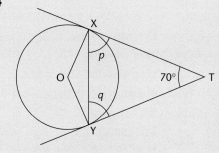

ACTIVITY 4

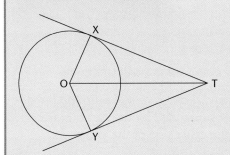

T is a point outside a circle, centre O.

TX and TY are the tangents from T to the circle.

Prove triangles OTX and OTY are congruent.

$XT^2 = OT^2 - OX^2$ (Pythagoras)

$YT^2 = OT^2 - OY^2$ (Pythagoras)

therefore $XT^2 = YT^2$ and $XT = YT$.

Activity 4 cont'd

Triangles OXT and OYT have corresponding sides equal and therefore corresponding angles equal.

OX = OY   (radii)

Angles OXT and OYT are 90°   (angle between radius and tangent)

TO is common

therefore triangle OTX is congruent to triangle OTY.   (RHS)

If the triangles OTX and OTY are congruent then

XT = YT; angle XTO = angle YTO; angle TOX = angle TOY

> Tangents drawn from a point to a circle are equal in length. They subtend equal angles at the centre of the circle and they make equal angles with the straight line joining the centre of the circle to the point.

ACTIVITY 5

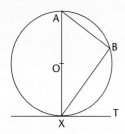

BX is a chord of the circle centre O.

TX is a tangent meeting the circle at X.

AX is a diameter.

Prove angle BXT = angle BAX.

(The solution is given below. Try to write out your own proof and then use this to check your work.)

Solution:

Angle ABX = 90°   (angle in a semi-circle)

Angle OXT = 90°   (angle between a diameter and a tangent)

therefore angle AXB = 90° − angle BXT

but angle AXB + angle BAX + angle ABX = 180°   (angle sum of triangle)

i.e. angle AXB + angle BAX = 90°

therefore 90° − angle BXT + angle BAX = 90°

therefore angle BXT = angle BAX.

This is a general result since angle XCB = angle XAB (subtended by arc XB).

> The angle between a tangent and a chord drawn from the point of contact is equal to the angle subtended by the chord in the alternate segment, that is on the other side of the chord.

EXERCISE 12.5A

1 In the following questions find the angles marked with letters. Gives reasons for your answers.

a)

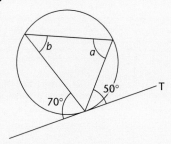

b)

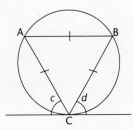

c)

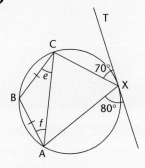

d)

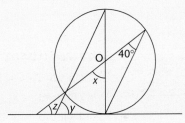

2

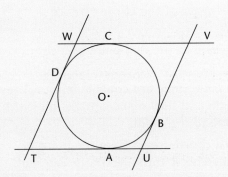

The sides of the quadrilateral TUVW are all tangents to the circle centre O.

Prove TU + WV = UV + WT.

EXERCISE 12.5B

In the following questions find the angles marked with letters. Give reasons for your answers.

1

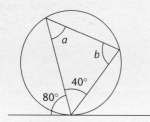

2

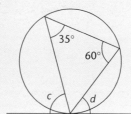

3

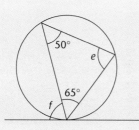

4

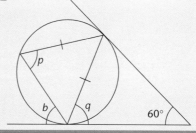

Key ideas

- The angle at the centre of a circle = twice the angle at the circumference subtended by the same arc (or same chord).

- The straight line which joins the centre of a circle to the mid-point of a chord is at right angles to the chord.

- Tangents drawn from the same point to the same circle are equal in length. They subtend equal angles at the centre of the circle and they make equal angles with the straight line joining the centre of the circle to the point.

- The angle between a tangent to a circle and a chord drawn from the point of contact is equal to the angle subtended by the chord in the alternate segment.

13 Using graphs to solve equations

You should already know

- how to use your calculator with trigonometrical functions and exponential functions
- Pythagoras' theorem.

EXAMPLE 1

Solve the simultaneous equations

$$y = x^2 + 3x - 7 \qquad [1]$$

and $\quad y = x - 4 \qquad\qquad [2]$

graphically using values of x from $^-5$ to $+ 2$.

$y = x^2 + 3x - 7$

x	$^-5$	$^-4$	$^-3$	$^-2$	$^-1$	0	1	2
x^2	25	16	9	4	1	0	1	4
$+3x$	$^-15$	$^-12$	$^-9$	$^-6$	$^-3$	0	3	6
$^-7$	$^-7$	$^-7$	$^-7$	$^-7$	$^-7$	$^-7$	$^-7$	$^-7$
$y = x^2 + 3x - 7$	3	$^-3$	$^-7$	$^-9$	$^-9$	$^-7$	$^-3$	3

$y = x - 4$

x	$^-5$	0	2
y	$^-9$	$^-4$	$^-2$

Example 1 cont'd

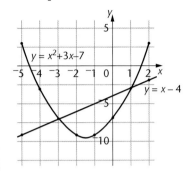

The curve and the line cross at $x = 1$, $y = {}^-3$ and $x = {}^-3$, $y = {}^-7$.

1 a) Draw the graph of $y = x^2 - 5x + 5$ for $x = {}^-2$ to $+5$.

   b) On the same grid draw the graph of $2x + y = 9$.

   c) Write down the coordinates of the points where the curve and line cross.

2 a) Draw the graph of $y = x^2 - 3x - 1$ for $x = {}^-4$ to $+3$.

   b) On the same grid draw the graph of $4x + y = 5$.

   c) Write down the coordinates of the points where the curve and line cross.

1 a) On one grid draw the graphs of $y = x^2 + 3$ and $y = 3x + 7$ for $x = {}^-2$ to $+5$.

   b) Find the simultaneous solutions of the two equations from your graph.

   c) Write down the coordinates of the points where the curve and line cross.

2 a) Draw the graph of $y = x^2 - 5x + 3$ for $x = {}^-2$ to $+4$.

   b) On the same grid draw the graph of $7x + 2y = 11$.

   c) Write down the coordinates of the points where the curve and line cross.

Using graphs to solve harder equations

Earlier, you saw how to solve equations like $x^2 - 5x + 3 = 0$ by drawing the graph of $y = x^2 - 5x + 3$ and reading off where it crossed the x-axis, where $y = 0$.

This can now be extended to include intersections with lines other than $y = 0$.

EXAMPLE 2

The graph of $y = x^2 - 3x + 1$ is drawn. (Do not draw it.)
How can you use the graph to find the solution of:

a) $x^2 - 3x + 1 = 0$

b) $x^2 - 3x - 1 = 0$?

a) This is where the curve crosses the line, $y = 0$.
Replace y by 0 in the equation.

b) $x^2 - 3x - 1 = 0$ is the same as $x^2 - 3x + 1 - 2 = 0$ or $x^2 - 3x + 1 = 2$.
Where $y = 2$ meets $y = x^2 - 3x + 1$ will give the solution.

EXAMPLE 3

a) Draw the graph of $y = x^2 - 1$ for $x = {}^-3$ to $+3$.

b) Use your graph to solve the equation $x^2 - 2 = 0$.

c) **(i)** Draw a suitable line so you can find the solution of $x^2 - x - 1 = 0$.

  (ii) Solve $x^2 - x - 1 = 0$ from your graph.

a)

x	$^-3$	$^-2$	$^-1$	0	1	2	3
x^2	9	4	1	0	1	4	9
$^-1$	$^-1$	$^-1$	$^-1$	$^-1$	$^-1$	$^-1$	$^-1$
$y = x^2 - 1$	8	3	0	$^-1$	0	3	8

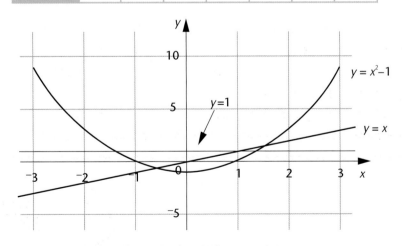

Example 3 cont'd

b) $x^2 - 2 = 0$ is the same as $x^2 - 1 - 1 = 0$ or $x^2 - 1 = 1$. Read where $y = 1$ meets $y = x^2 - 1$.

Solution: $x = {}^-1\cdot4$ or $1\cdot4$.

c) **(i)** $x^2 - x - 1 = 0$ is the same as $x^2 - 1 = x$. So the required line is $y = x$.

(ii) Solution is $x = {}^-0\cdot6$ or $1\cdot6$.

EXAMPLE 4

The graph of $y = x^3 - 5x$ is drawn.
(Do not draw the graph.)

What line needs to be drawn to find the solution of
$x^3 - 6x - 5 = 0$.

$x^3 - 6x - 5 = 0$ is the same as $x^3 - 5x - x - 5 = 0$
or $3 - 5x = x + 5$.

The line that needs to be drawn is $y = x + 5$.

Exam tip

When you are asked to solve an equation from your graph, make it clear what lines you are using to find the solution. Marks will usually be given for using the correct method even if the solution is not accurate.

Answers should normally be given to 1 decimal place unless there is a different instruction.

EXAMPLE 5

a) Draw the graph of $y = x^2 - 5x + 3$ for $x = {}^-1$ to $+6$.

b) From your graph solve the inequality $x^2 - 5x + 3 < 0$.

x	⁻1	0	1	2	3	4	5	6
x²	1	0	1	4	9	16	25	36
⁻5x	5	0	⁻5	⁻10	⁻15	⁻20	⁻25	⁻30
3	3	3	3	3	3	3	3	3
y = x² – 5x + 3	9	3	⁻1	⁻3	⁻3	⁻1	3	9

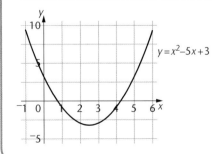

b) The solution is where the curve goes below the x-axis since
$y < 0$, giving $0\cdot7 < x < 4\cdot3$.

EXERCISE 13.2A

(Do not draw the graphs in questions 1 to 5.)

1 The graph of $y = x^2 - 6x + 4$ is drawn.

How can you use the graph to find the solution of:

a) $x^2 - 6x + 4 = 0$ **b)** $x^2 - 6x + 2 = 0$?

2 The graph of $y = 2x^2 - 3x$ is drawn.

How can you find the solution of $2x^2 - 3x - 5 = 0$?

3 The graphs of $y = x^2$ and $y = 4x - 3$ are drawn on the same grid. What is the equation whose solution is found at the intersection of the two graphs?

4 The graph of $y = x^3 - 4x$ is drawn. What other graph needs to be drawn to find the solution of $x^3 - x^2 - 4x + 3 = 0$ when they cross?

5 The intersection of two graphs is the solution to the equation $x^2 - 4x - 2 = 0$. One of the graphs is $y = x^2 - 5x + 1$. What is the other graph?

6 **a)** Draw the graph of $y = x^2 - 5$ for $x = ^-3$ to $+3$.

b) From your graph solve the equations:

(i) $x^2 - 5 = 0$ **(ii)** $x^2 - 3 = 0$.

7 **a)** Draw the graph of $y = 2x^2 - 3x$ for $x = ^-2$ to $+4$.

b) Use your graph to solve the equation $2x^2 - 3x - 5 = 0$.

8 **a)** Draw the graphs of $y = x^2 + 5$ and $y = 3x + 7$ for $x = ^-2$ to $+4$.

b) What is the equation of the points where they intersect?

c) Solve this equation from your graph.

9 **a)** Draw the graph of $y = x^3 - 3x$ for $x = ^-3$ to $+3$.

b) **(i)** Draw another graph so that the equation of the points of intersection is: $x^3 - 6x + 5 = 0$.

(ii) Use the graph to solve the equation.

10 a) Draw the graph of $x^2 - 2x - 4$ for $x = ^-2$ to $+4$.

b) Find the solutions of the following from the graph.

(i) $x^2 - 2x - 7 = 0$ **(ii)** $x^2 - 4x - 6 = 0$ **(iii)** $x^2 - 2x - 4 > 0$.

EXERCISE 13.2B

(Do not draw the graphs in questions 1 to 5.)

1 The graph of $y = x^2 - 8x + 2$ is drawn. How can you use the graph to find the solution of:

a) $x^2 - 8x + 2 = 0$ **b)** $x^2 - 8x + 6 = 0$.

2 The graph of $y = 2x^2 - x - 2$ is drawn. How can you find the solution of $2x^2 - x - 5 = 0$?

3 The graphs of $y = x^2 + 3x$ and $y = 4x - 3$ are drawn on the same grid. What is the equation whose solution is found at the intersection of the two graphs?

4 The graph of $y = x^3 - 2x^2$ is drawn. What other graph needs to be drawn to find the solution of $x^3 - x^2 - 4x + 3 = 0$ when they cross?

5 The intersection of two graphs is the solution to the equation $x^2 - 5x - 3 = 0$. One of the graphs is $y = x^2 - x + 1$. What is the other graph?

6 **a)** Draw the graph of $y = 2x^2 - 10$ for $x = {}^-3$ to $+3$.

b) From your graph solve the equations:

(i) $2x^2 - 10 = 0$ **(ii)** $2x^2 - 3 = 0$.

7 **a)** Draw the graphs of $y = x^2 + 2$ and $y = 2x + 7$ for $x = {}^-2$ to $+4$.

b) What is the equation of the points where they intersect?

c) Solve this equation from your graph.

8 **a)** Draw the graph of $y = x^2 - 5x$ for $x = {}^-2$ to $+4$.

b) (i) Draw another line so that the intersection is $x^2 - 3x - 3 = 0$.

(ii) Solve this equation from the graph.

9 **a)** Draw the graph of $y = x^3 - 5x$ for $x = {}^-3$ to $+3$.

b) (i) Draw another graph so that the equation of their points of intersection is:
$x^3 - x^2 - 5x + 5 = 0$.

(ii) Use the graph to solve the equation.

10 a) Draw the graph of $x^2 + 5x + 4$ for $x = {}^-6$ to $+1$.

b) Find the solutions of the following from the graph:

(i) $x^2 + 5x + 4 = 0$ **(ii)** $x^2 + 3x - 3 = 0$ **(iii)** $x^2 + 5x + 2 < 0$.

Graphs of exponential growth and decay functions

EXAMPLE 6

The number of bacteria present doubles every hour.

The table of values for the amount of bacteria is:

Number of hours	0	1	2	3	4
Number of bacteria	500	1000	2000	4000	8000

The graph is:

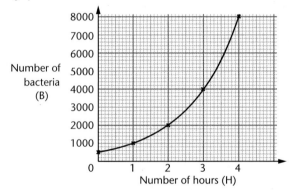

The shape of this graph is typical of an exponential growth function.

Its equation will be: $B = 500 \times 2^H$

500 because this is the initial value.

2 because the number of bacteria doubles each time.

EXAMPLE 7

The population of a certain species of bird is dropping by 20% every 10 years.

The table of values is:

Year	1970	1980	1990	2000	2010	2020
Number of birds	50 000	40 000	32 000	25 600	20 480	16 384

The graph is:

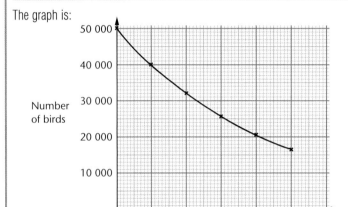

The shape of this graph is typical of an exponential decay function.

EXAMPLE 8

Plot a graph of $y = 3^x$ for values of x from $^-2$ to $+3$.

Use your graph to estimate:

a) the value of y when $x = 2·4$

b) the solution to the equation $3^x = 20$.

The table of values is:

x	$^-2$	$^-1$	0	1	2	3
y	0·111	0·333	1	3	9	27

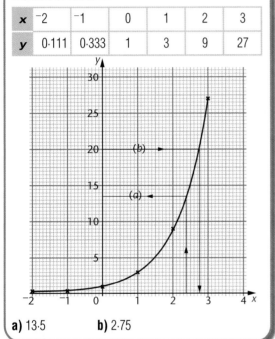

a) 13·5 **b)** 2·75

EXAMPLE 9

Plot a graph of $y = 2^{-x}$ for values of x from $^-4$ to $+2$.

Use your graph to estimate:

a) the value of y when $x = 0·5$

b) the solution to the equation $2^{-x} = 10$.

x	$^-4$	$^-3$	$^-2$	$^-1$	0	1	2
y	16	8	4	2	1	0.5	0.25

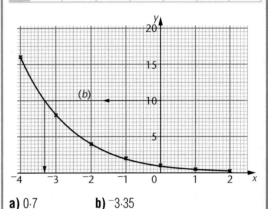

a) 0·7 **b)** $^-3·35$

Note that, although the scales make it difficult to show here, neither graph ever reaches the x-axis.

EXERCISE 13.3A

1 Two people start a rumour that goes round a school.

The following tables shows the number of people (n) who have heard the rumour after t hours.

Time (t) in hours	0	1	2	3	4	5
Number of people (n)	2	6	18	54	162	486

Find a formula for n in terms of t.

Chapter 13 *Using graphs to solve equations*

Exercise 13.3A cont'd

2 Plot a graph of $y = 2^x$ for values of x from $^-2$ to 5. Use a scale of 2 cm to 1 unit on the x-axis and 2 cm to 5 units on the y-axis.

Use your graph to estimate:

a) the value of y when $x = 3\cdot2$

b) the solution to the equation $2x = 20$.

3 Copy and complete the table of values for $y = 3^{-x}$.

x	0	0·5	1	1·5	2	2·5	3	3·5	4
y	1								

Plot the graph of $y = 3^{-x}$ for these values. Use a scale of 2 cm to 1 unit on the x-axis and 1 cm to 0·1 units on the y-axis.

Use your graph to estimate:

a) the value of y when $x = 1\cdot2$

b) the solution to the equation $3^{-x} = 0\cdot1$.

EXERCISE 13.3B

1 The mass (m) of a chemical present after t minutes during a chemical reaction is given in the table below.

Time (t) in minutes	0	1	2	3	4	5
Mass in grams (m)	100	50	25	12·5	6·25	3·125

Find a formula for m in terms of t.

2 Plot a graph of $y = 1\cdot5^x$ for values of x from $^-3$ to +5. Use a scale of 2 cm to 1 unit on both axes.

Use your graph to estimate:

a) the value of y when $x = 2\cdot4$ **b)** the solution to the equation $1\cdot5^x = 6$.

3 Copy and complete the table of values for $y = 4^{-x}$.

x	$^-2\cdot5$	$^-2$	$^-1\cdot5$	$^-1$	0	0·5	1
y		16					

Plot the graph of $y = 4^{-x}$ for these values. Use a scale of 2 cm to 1 unit on the x-axis and 2 cm to 5 units on the y-axis.

Use your graph to estimate:

a) the value of y when $x = ^-1\cdot8$ **b)** the solution to the equation $4^{-x} = 25$.

The equation of a circle

Draw the graph of $x^2 + y^2 = 25$.

Remember Pythagoras' theorem. It looks like this equation. Can you spot a pair of values for x and y which fit? What about (3, 4)? (4, 3)? (5, 0). These all satisfy the equation.

So do: (⁻5, 0), (⁻4, 3), (⁻3, 4), (0, 5), (3, 4), (4, 3) (5, 0), (4, ⁻3), (3, ⁻4), (0, ⁻5), (⁻3, ⁻4), (⁻4, ⁻3).

Can you see why the graph is a circle?

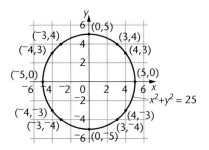

The general equation of a circle centre (0,0) and radius r is $x^2 + y^2 = r^2$, as shown by Pythagoras' theorem.

Exam tip

Working out the values in the usual way may cause problems!

EXAMPLE 10

Draw the graph of $x^2 + y^2 = 9$.

This is the graph of the circle centre (0, 0) and radius 3. The only integer values it passes through are (⁻3, 0) (0, 3), (3, 0), (0, ⁻3).

When $x = 2$ or ⁻2, $4 + y^2 = 9$

$$y^2 = 5$$

$$y = \pm\sqrt{5} = +2.24 \text{ or } ⁻2.24.$$

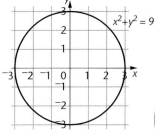

All these 8 points can be plotted and the circle will pass through them.

Exam tip

It is easier to find the radius and use compasses rather than to calculate points.

EXAMPLE 11

a) On the same grid draw the graphs of $x^2 + y^2 = 16$ and $y = x + 2$.

b) Use the graph to solve the two equations simultaneously, giving the answers correct to 1 decimal place.

a)

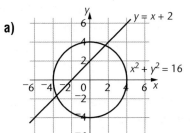

b) The solution is where the two graphs cross (⁻3·6, ⁻1·6) or (1·6, 3·6).

Chapter 13 *Using graphs to solve equations*

EXERCISE 13.4A

1 Draw the graph of $x^2 + y^2 = 64$. Use a scale of 1 cm to 2 units for both x and y.

2 **a)** Draw the graphs of $x^2 + y^2 = 100$ and $y = x + 2$ on the same grid. Use a scale of 1 cm to 2 units for both x and y.

   b) Find the coordinate of the points where the two graphs cross.

3 **a)** Draw the graphs of $x^2 + y^2 = 64$ and $y = 2x + 8$ on the same grid. Use a scale of 1 cm to 2 units for both x and y.

   b) Use the graph to solve simultaneously the equations $x^2 + y^2 = 64$ and $y = 2x + 8$. Give the answers either as whole numbers or correct to 1 decimal place.

EXERCISE 13.4B

1 Draw the graph of $x^2 + y^2 = 625$. Use a scale of 1 cm to 5 units for both x and y.

2 **a)** Draw the graphs of $x^2 + y^2 = 25$ and $y + x = 7$ on the same grid. Use a scale of 1 cm to 2 units for both x and y.

   b) Find the coordinates of the points were the two graphs meet.

3 **a)** Draw the graphs of $x^2 + y^2 = 49$ and $y = x + 5$ on the same grid. Use a scale of 1 cm to 2 units for both x and y.

   b) Use the graph to solve simultaneously the equations $x^2 + y^2 = 49$ and $y = x + 5$.

Give the answers either as whole numbers or correct to 1 decimal place.

Key ideas

- Exponential growth and decay functions have an equation of the form $y = Ab^x$ for growth or $y = Ab^{-x}$ for decay where A is the initial value and b is the growth/decay rate.

- The equation of a circle centre (0,0) and radius r is $x^2 + y^2 = r^2$.

- When solving simultaneous equations graphically, draw the two graphs and find where they intersect.

- Sometimes, when solving an equation graphically, you will need to adapt the equation. Rearrange it so that the equation of the graph drawn is on the left of the equals sign. What appears on the right is the graph that needs to be drawn.

Chapter 13 *Using graphs to solve equations*

Revision exercise

1 Draw a histogram to show the following distribution of the weights, to the nearest kilogram, of 50 Year 9 pupils.

Weight (W kg)	Frequency (f)
$32 \leq W < 34$	1
$34 \leq W < 39$	5
$39 \leq W < 43$	7
$43 \leq W < 47$	8
$47 \leq W < 51$	14
$51 \leq W < 59$	9
$59 \leq W < 70$	6

2 This histogram shows the masses, in grams, of plums picked in an orchard.

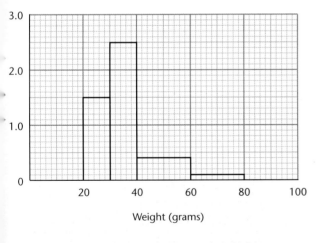

Weight (grams)

How many plums were picked?

3 The heights of students in two classes are measured. The results are given in the table below:

Class 11A	
Height (H cm)	Frequency (f)
$130 \leq H < 140$	1
$140 \leq H < 150$	4
$150 \leq H < 160$	9
$160 \leq H < 170$	8
$170 \leq H < 180$	2
$180 \leq H < 190$	2

Class 11B	
Height (H cm)	Frequency (f)
$120 \leq H < 130$	4
$130 \leq H < 140$	5
$140 \leq H < 150$	8
$150 \leq H < 160$	3
$160 \leq H < 170$	3
$170 \leq H < 180$	1

Show the data on two histograms.

4 In the following questions O is the centre of each circle. In each question calculate the size of the angles marked with letters.

a)

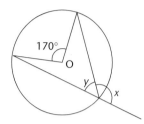

b)

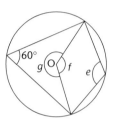

c)

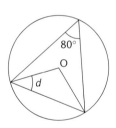

d)

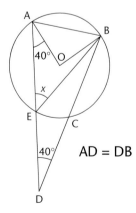

AD = DB

e)

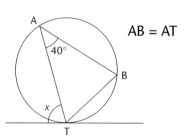

AB = AT

f)

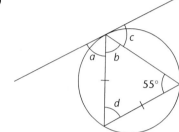

g)

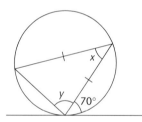

5 a) Draw the graphs of
$y = x^2 - 2x + 3$ and $y = 4x + 1$, on the same grid.

Use values of x from 0 to 6.

b) Use the graphs to solve, simultaneously, the equations $y = x^2 - 2x + 3$ and $y = 4x + 1$.

Give the answers correct to 1 decimal place.

6 a) Draw the graphs of
$x^2 + y^2 = 9$ and $y = x + 2$ on the same grid.

Use a scale of 1 cm to 1 unit for both x and y.

b) Use the graph to solve, simultaneously, the equations
$x^2 + y^2 = 9$ and $y = x + 2$.

Give the answers correct to 1 decimal place.

7 Solve these equations simultaneously.
$x^2 + y^2 = 36$ and $y = x + 6$.

8 a) Draw the graph of $y = x^2 + 3x - 7$ for $x = {}^-6$ to $+3$.

b) Use the graph to solve:

(i) $x^2 + 3x - 7 = 0$

(ii) $x^2 + 3x - 10 = 0$.

c) (i) Find the line that must be drawn to solve
$x^2 + x - 4 = 0$.

(ii) Draw the line and use it to solve
$x^2 + x - 4 = 0$.

9 Copy and complete the table of values for $y = 2^{-x}$.

x	0	0·5	1	1·5	2	2·5	3	3·5	4
y	1								

Plot the graph of $y = 2^{-x}$ for these values. Use a scale of 2 cm to 1 unit on the x-axis and 1 cm to 0·1 units on the y-axis.

Use your graph to estimate:

a) the value of y when $x = 1·8$

b) the solution to the equation
$2^{-x} = 0·6$.

Stage 10

Chapter 1	**Growth and decay**	**121**
	Exponential growth	121
	Exponential decay	122
	Using trial and improvement to solve growth and decay problems	123
Chapter 2	**Finding formulae that connect data**	**128**
	General laws in symbolic form	128
	Sequences	130
	Approximation to linear graphs	133
	Non-linear graphs	134
Chapter 3	**Surveys and sampling**	**140**
	Surveys	141
	Questionnaire design	141
	Sampling	142
	Bias	146
	Stratified random sampling	146
Chapter 4	**Rational and irrational numbers**	**153**
	Rational numbers	153
	Irrational numbers	153
	Recurring decimals and fractions	154
	Simplifying surds	156
	Manipulation of numbers like $a + b\sqrt{c}$	157
	Rationalising denominators	158
Revision exercise A1		**162**
Chapter 5	**Trigonometry in non-right-angled triangles**	**164**
	Notation	164
	The rules	164
	General formula for the area of any triangle	175
Chapter 6	**Trends and time series**	**177**
Chapter 7	**Congruency – proving and using**	**183**
	Congruent triangles	183
Chapter 8	**Calculating roots of equations**	**189**
	Completing the square	190
	Using the quadratic formula	191
Revision exercise B1		**195**

CONTENTS

1 Growth and decay

You should already know

- how to deal with negative powers
- how to draw graphs of exponential growth and decay.

Exponential growth

£200 is invested at 5% per year compound interest.

After 1 year the investment will be worth $200 \times 1.05 = £210$.

After 2 years the investment will be worth $210 \times 1.05 = £220.50$.

After 3 years the investment will be worth $£220.50 \times 1.05 = £231.53$.

This sort of calculation, involving a constant multiplier, is an example of **exponential growth**. If the multiplier is a larger number, the figures get very large, very quickly.

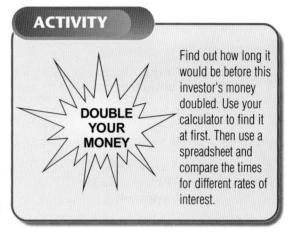

ACTIVITY

DOUBLE YOUR MONEY

Find out how long it would be before this investor's money doubled. Use your calculator to find it at first. Then use a spreadsheet and compare the times for different rates of interest.

To do the calculation in the same way for a large number of years is time consuming. It would be easier, if it was possible, to find a formula for the calculation.

Look again at the calculations for two and three years.

Another way of looking at the calculation for two years is $200 \times 1.05 \times 1.05 = 200 \times 1.05^2$. Similarly, the calculation for three years is $200 \times 1.05 \times 1.05 \times 1.05 = 200 \times 1.05^3$.

So, after 20 years, the investment will be worth 200×1.05^{20}.

On a calculator this calculation is done using the power button. This is usually labelled $\boxed{y^x}$ $\boxed{x^y}$ or, on many modern calculators, $\boxed{\wedge}$.

The calculation is then simply $200 \times 1.05 \boxed{\wedge} 20 = £530.66$.

The formula for this calculation is $A = 200 \times 1.05^n$, where A is the amount the investment is worth and n is the number of years.

EXAMPLE 1

The number of bacteria present doubles every hour. If there were 500 present at 12 noon, find the number present:

a) at 2 p.m. **b)** 3 p.m.

c) at midnight **d)** after n hours.

a) $500 \times 2 \times 2 = 2000$.

b) $500 \times 2^3 = 4000$ or $2000 \times 2 = 4000$.

c) $500 \times 2^{12} = 2\,048\,000$.

d) number present $= 500 \times 2^n$.

Exponential decay

A car depreciates in value by 15% per year. It cost £12 000 when it was new.

After 1 year it will be worth $12\,000 \times 0.85 = £10\,200$.

After 2 years it will be worth $10\,200 \times 0.85 = £8670$.

After 3 years it will be worth $8670 \times 0.85 = £7369.50$. And so on.

This calculation, where the constant multiplier is less than 1, is an example of **exponential decay**.

The calculations involved work exactly like the ones for exponential growth, except that the multiplier is less than 1 and so the numbers get smaller instead of bigger.

After 10 years the car will be worth $12\,000 \times 0.85^{10} = £2362.49$.

The formula for this calculation is $A = 12\,000 \times 0.85^n$,

where £A is the amount the car is worth and n is the number of years.

EXAMPLE 2

The population of a certain species of bird is dropping by 20% every 10 years. If there were 50 000 in 1970, how many will there be:

a) in 2010 **b)** 2020 **c)** 2100 **d)** n years after 1970?

a) $\dfrac{2010 - 1970}{10} = 4$ $50\,000 \times 0.8^4 = 20\,480$.

b) $20\,480 \times 0.8 = 16\,384$ (or $50\,000 \times 0.8^5 = 16\,384$).

c) $\dfrac{2100 - 1970}{10} = 13$ $50\,000 \times 0.8^{13} = 2749$.

d) Since the population decreases by 10% every 10 years, the number will be $50\,000 \times 0.8^{\frac{n}{10}}$.

The calculations in Examples 1 and 2 show the dramatic changes that exponential growth and decay calculations can make.

The same effect of exponential decay can be achieved using a **negative power**.

For example, if a population starts at 1 million and is halved every year:

after 5 years the population is $1\,000\,000 \times 0.5^5 = 31\,250$

after n years the population is $1\,000\,000 \times 0.5^n$.

Since $0.5 = \frac{1}{2} = 2^{-1}$, the calculation can be written:

after 5 years the population is $1\,000\,000 \times 2^{-5} = 31250$

after n years the population is $1\,000\,000 \times 2^{-n}$.

Using trial and improvement to solve growth and decay problems

Look again at the bacteria data in Example 1.

In Stage 9, a graph was drawn to show these data. This graph could be used to answer questions such as:

How many hours after 12 noon is the population 6000?

From the graph, as shown, the answer is approximately 3.6. A more accurate answer may be obtained using **trial and improvement**. This requires use of the power key on the calculator.

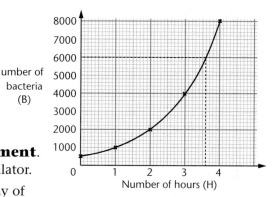

The following table demonstrates a possible way of obtaining the solution correct to 2 decimal places.

Number of hours (H)	Number of bacteria $= 500 \times 2^H$	Decision
3.6	6062.86…	Too large
3.5	5656.85…	Too small but further away. Try between 3.5 and 3.6 but near 3.6.
3.59	6020.987…	Too large, so try smaller.
3.58	5979.397	Too small, so try halfway between.
3.585	6000.156	Solution is between 3.58 and 3.585, so is **3.58** correct to 2 decimal places.

EXAMPLE 3

Solve $2^{-x} = 0.6$. Use trial and improvement and give your solution correct to 1 decimal place.

x	2^{-x}	Decision
0	1	
1	0.5	Solution is between 0 and 1, try nearer to 1.
0.9	0.535…	
0.8	0.574…	
0.7	0.615…	Solution is between 0.7 and 0.8. Try 0.75.
0.75	0.594…	Solution is between 0.7 and 0.75 so is **0.7** to 1 d.p.

EXERCISE 1.1A

1 **a)** Complete the following table for a sum of money £y growing at 12% each year for x years.

x	0	1	2	3	4	5
y	500	560	627.20			

   b) Using trial and improvement, or otherwise, solve the equation
   $500 \times 1.12^x = 750$, giving the value of x correct to 1 d.p.

2 **a)** Sketch the graph of $y = 2^x$ for $x = 0$ to 6.

   b) Use trial and improvement, correct to 1 d.p. to find the value of x when $y = 200$.

3 The size, y, of a population of bacteria is growing according to the rule
   $y = 25 \times 1.02^t$, where t minutes is the measured time.

   a) How many bacteria are there at time $t = 0$?

   b) What will the population be 5 hours after starting to measure the time?

   c) Find how long it took the population to double in size.

4

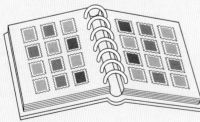

In 2000, the value of Bharat's stamp collection was £85. Assume the value increases at 5% each year.

   a) What will its value be in 2004?

   b) Write, as simply as possible, an expression for its value x years after 2000.

   c) Find by calculation how many years it will take for the value of the stamp collection to double.

Exercise 1.1A cont'd

5  A population of bats is declining at a rate of 15% each year. At the start of 2000 there were 140 bats. How many years after the start of 2000 would the population be 80 bats? Give your answer to 1 d.p.

6 A radioactive element has a mass of 50 g. Decay reduces its mass by 10% each year. Use trial and improvement to estimate the time taken for its mass to halve. Give your answer in years to 1 d.p.

7 £5000 is invested at 3% compound interest per year.

    a) Calculate the value of the investment after 20 years.

    b) Find, to the nearest year, the time taken for the investment to be worth £12 000.

8 A car cost £16 000 when new and depreciates in value by 16% each year.

    a) State a formula for the value of the car after n years.

    b) Find, in years to 1 d.p., the age of the car when its value is £5000.

9 The graph shows a function of the form $y = ab^x$. Find the values of a and b.

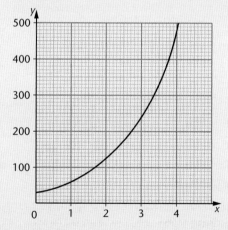

10 Use trial and improvement to solve these equations, giving your solution correct to 2 d.p.

    a) $2^x = 20$ **b)** $3^{-x} = 0.5$

EXERCISE 1.1B

1 **a)** Complete this table for the function $y = 3^x$.

x	0	1	2	3	4	5
y						

   b) Find the value of y when $x = 12$.

   c) Using trial and improvement, or otherwise, find, correct to 1 d.p., the value of x when $y = 6\,000$.

2 The number of bacteria in a colony doubles every 15 minutes. At the start of measuring time, there were 50 bacteria.

   a) Complete the table to show the number of bacteria during the first two hours.

Time (hours)	0	0·25	0·5	0·75	1·0	1·25	1·5	1·75	2·0
Number of bacteria	50								

   b) Calculate the population after 6 hours. Give your answer in standard form, to 2 s.f. (significant figures).

   c) Find how long it took for the population to reach 4 000. Give your answer in hours, correct to 1 d.p.

3 The value of an investment is declining at a rate of 10% per year. Initially the investment was worth £2000.

   a) Calculate how much it is worth after 4 years.

   b) Calculate, by trial and improvement, or otherwise, how long after the start the investment will be worth £1000, giving your answer in years correct to 1 d.p.

4 A population of bacteria is decreasing at a rate of 14% per hour. Initially it is estimated that there are 2000 bacteria.

   a) Sketch a graph to show this population in the first 6 hours.

   b) Find, correct to 1 d.p., the number of hours after which the population is 750.

5 The value of an antique increases by 10% each year. At the beginning of 2001 it was worth £450. How many years later will it be worth £2000? Give your answer correct to 1 d.p.

6 A car cost £9000 when new. Its value depreciates by 12% each year. Find, correct to 1 d.p., how many years later it will be worth £3000.

Exercise 1.1B cont'd

7 £2000 is invested at 6% compound interest.

    a) Write a formula for the value of this investment n years later.

    b) Find, to the nearest year, the time taken for the investment to be worth £6000.

8 The mass (m grams) of a chemical present after t minutes during a chemical reaction is given by the formula $m = 100 \times 0.5^t$.

    a) What mass was present initially?

    b) Find, correct to 2 d.p., the value of t when the mass was 20 g.

9 The curve $y = ab^x$ passes through the points (0, 5) and (3, 20·48). Find the values of a and b.

10 Use trial and improvement to solve these equations, giving your solutions correct to 2 d.p.

    a) $1.5^x = 6$ **b)** $4^{-x} = 0.1$ **c)** $6^x = 0.03$

Key ideas

- When a value is multiplied by a constant multiplier (greater than one), this is an example of exponential growth.
- When a value is multiplied by a constant multiplier (less than one), this is an example of exponential decay.
- Exponential functions are functions of the form $y = ab^x$. Since $b^0 = 1$, the graphs of these functions always go through (0, a).
- To solve an exponential equation when the power is unknown, use the power key and trial and improvement.

2 Finding formulae that connect data

You should already know

- the equation of a straight line is $y = mx + c$, when m is the gradient and c is the y intercept.

General laws in symbolic form

In previous modules you have seen how to recognise sequences and how to write the nth term in terms of n. In this chapter this idea will be extended and other algebraic formulae will be expressed.

The most important thing, before starting to write down an algebraic formula, is to define carefully the letters you are using to stand for quantities and, if units are involved, what they are.

EXAMPLE 1

Write down a formula for the cost of a pens and b notebooks.

Let p be the cost of a pen in pence.

Let n be the cost of a notebook in pence.

Let T be the total cost in pence.

The formula is then $T = ap + bn$.

Exam tip

It is very common to see undefined variables being used. Even when the attempt is made it is often very vague. For example, p = pens does not tell us that p = the cost of a pen or what units are being used.

The next important step is to make sure that, if you have spotted the rule, you write it down accurately using correct algebra.

EXAMPLE 2

Write down a formula for the mean of three numbers.

Let the three numbers be a, b and c.

Let the mean $= m$.

The formula is $m = \dfrac{a + b + c}{3}$

> Note that it is usual to use fraction lines rather than the ÷ sign in algebraic formulae. Even if you do use the ÷ sign it is important to use brackets to make sure the order of operations is correct.
>
> $m = a + b + c ÷ 3$ is not correct as it would mean that c was divided by 3 first and then added to $a + b$.
>
> $m = (a + b + c) ÷ 3$, whilst mathematically correct, is not the usual way to write a formula.
>
> $m = \dfrac{1}{3}(a + b + c)$ is an acceptable algebraic formula.

Some formulae use both the capital (upper case) version of a letter and the small (lower case) version as well. For this reason it is important to keep to the version of the letter as defined.

For example, the formula for the area of the shaded section in this diagram is

$A = \pi R^2 - \pi r^2$

where A = the area in cm^2

R = the radius of the large circle in cm

r = the radius of the small circle in cm.

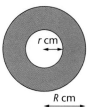

ACTIVITY

Write the shaded areas as a formula.

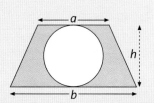
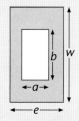
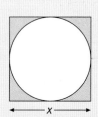

Chapter 2   *Finding formulae that connect data*

Sequences

EXAMPLE 3

Find a formula for the nth term of this sequence
2, 6, 18, 54, 162 ...

> When you are answering a question like this it is usually best to produce a table.

Let U_n = the nth term

n	1	2	3	4	5
U_n	2	6	18	54	162

> It is fairly easy to spot a 'term to term' rule as each term is three times the previous one.

This is written as $U_{n+1} = 3 \times U_n$ and, together with $U_1 = 2$, defines the sequence.

> To get, for example, the 20th term, it is necessary to get the 19th first, so an algebraic rule would be better.

The 2nd term = 2×3
The 3rd term = $2 \times 3 \times 3 = 2 \times 3^2$
The 4th term = $2 \times 3 \times 3 \times 3 = 2 \times 3^3$
The 5th term = $2 \times 3 \times 3 \times 3 \times 3 = 2 \times 3^4$
The nth term = $2 \times 3^{n-1}$
So $U_n = 2 \times 3^{n-1}$
$U_{20} = 2 \times 3^{19} = 2\ 324\ 522\ 934.$

> This sequence, where each term is multiplied by a fixed number to get the next, is called a **geometric sequence**.

EXAMPLE 4

Find a formula for the nth triangle number. Use the formula to find the 40th triangle number.

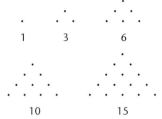

Let T_n be the nth triangle number.

> The term to term rule is fairly easy. For the 4th, you add 4 to the third, for the fifth you add 5 to the 4th and so on, so $T_n = T_{n-1} + n$ but the algebraic rule is somewhat harder.

n	1	2	3	4	5
T_n	1	3	6	10	15

> Even when the numbers are put into a table the formula is difficult to spot. Add another row to the table for $n + 1$.

n	1	2	3	4	5
$n + 1$	2	3	4	5	6
T_n	1	3	6	10	15

Even now it is not easy, but notice that

$$3 = \frac{2 \times 3}{2}, \quad 6 = \frac{3 \times 4}{2}, \quad 10 = \frac{4 \times 5}{2} \ldots$$

so $$T_n = \frac{n(n+1)}{2}$$

$$T_{40} = \frac{40 \times 41}{2} = 20 \times 41 = 820$$

This formula is worth remembering as triangle numbers arise quite often in investigation work.

EXERCISE 2.1A

1 A factory makes chairs and tables. A chair takes 5 hours to make and a table takes 4 hours. Find a formula for the total time spent making chairs and tables in a week.

2 Minibuses carry 10 passengers and taxis carry 4 passengers. Find a formula for the total number of people carried by taxis and minibuses.

> **Exam tip**
>
> In questions where you choose your own letters, make sure you define them carefully.

3 To find the geometric mean of two numbers, you multiply the numbers together and find the square root of the answer. Write this relationship as a formula.

4 A two-digit number has the first digit a and the second digit b. Write down a formula for the number.

5 A ferry carries cars and lorries. On average a car takes up $20\,m^2$ of space and a lorry takes up $50\,m^2$ space. The decks have a total space of $5500\,m^2$. Write down an inequality satisfied by the number of cars and lorries.

6 Here is a sequence.

4, 20, 100, 500, 2500 …

Find:

a) the nth term. **b)** the 10th term.

7 Here is a sequence.

128, 64, 32, 16, 8 …

Find:

a) the nth term **b)** the 10th term.

8 When a stone is dropped down a well, the distance it falls is proportional to the square of the time for which it has been falling. Write this relationship algebraically

a) using the symbol $\propto$ **b)** as a formula using the constant k.

9 Peter invests £P at 5% per annum compound interest. Write down a formula for the amount the investment is worth after:

a) 1 year **b)** 3 years **c)** n years.

10 The coordinates of P are (a, b). The coordinates of Q are (c, d). Find a formula for the distance PQ.

EXERCISE 2.1B

1 An electricity bill is made up of a fixed charge plus a charge for each unit of electricity used. Write a formula for the total bill.

2 Jane buys some books for £3 each. She sells some of them for £5 each. To get rid of the remainder she sells them at £2 each. Find, and simplify, a formula for her profit.

3 Asif buys some cans of cola at 50p each and some packets of crisps at 30p each. He has £5. Write down an inequality satisfied by the number of cans of cola and number of packets of crisps he buys.

4 A three-digit number has the first digit a, the second digit b and the third digit c. Write down a formula for the number.

5 To find the volume of a pyramid, multiply the area of the base by the height and divide by 3. Write a formula for the volume of a pyramid with a base that is
   a) square, **b)** a rectangle.

6 Here is a sequence.

   0·5, 2, 8, 32, 128 …

   Find:

   a) the nth term **b)** the 10th term.

7 Here is a sequence.

   2, 5, 10, 17, 26 …

   Find:

   a) the 7th term **b)** the nth term **c)** the 20th term.

8 Newton's Law of Gravitation states that the force of attraction between two bodies is inversely proportional to the square of the distance between them.

   Write this relationship algebraically

   a) using the symbol $\propto$ **b)** as a formula using the constant k.

9 The value of a car depreciates by 15% per year. If the car was worth £C when new, write a formula for the value after

   a) 1 year **b)** 3 years **c)** n years.

10 Write down a formula for the gradient of the line PQ in terms of the coordinates of P and Q.

11 The total stopping distance for a car consists of two parts: a thinking distance, which is proportional to the speed, plus a braking distance that is proportional to the square of the speed.

   Using two constants, write down a formula for the total stopping distance in terms of the speed.

> **Exam tip**
>
> In questions where you choose your own letters, make sure you define them carefully.

Approximation to linear graphs

In scientific experiments you will often have plotted graphs of your results.

For example, if you hang weights on a light spring you can measure the extension in the spring and plot the tension against the extension. The table may look like this.

Tension in Newtons (T)	1	2	3	4	5
Extension in mm (e)	9·2	17·3	26·1	37·1	43·8

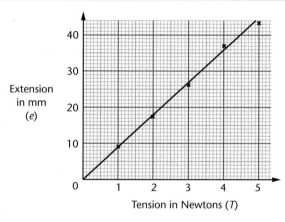

These results lead to the graph shown.

The points are not exactly in a straight line but, since the figures came from an experiment, it seems likely that the true relationship is linear and the variations are caused by experimental error.

If the points lie in a sufficiently good straight line to make you believe the relationship is linear, then you can draw the best straight line you can through the points. This has been done on the diagram.

This is similar to the lines of best fit you drew for correlation graphs. In the same way, the line should reflect the slope of the points and have approximately the same number of points on either side. If one point lies a long way from the straight line then the experiment should be repeated for that reading.

When you drew correlation graphs it was quite common for the line of best fit not to go through the origin. In experimental situations the real situation should be examined to see if it is expected that the graph should go through the origin. In the above case it seems logical that if there is no tension there will be no extension and so the line will go through the origin.

Once the line of best fit is drawn you can find its equation. This will give you the relationship between the two quantities you have graphed. The two quantities that you need to read off from the graph are gradient and y intercept. These can be substituted into $y = mx + c$ to find the relationship.

In the above case $c = 0$ and $m = \dfrac{45}{5} = 9$.

So the relationship is $e = 9T$.

> **Exam tip**
>
> Make sure you read off the scales when finding the gradient. Do not use squares.

Chapter 2   *Finding formulae that connect data*

In writing the relationship, it is particularly important that the variables are well defined, together with their units.

Here T is the tension in Newtons. e is the extension in mm.

Non-linear graphs

When the relationship is non-linear, it is a little more difficult to find.

The table below shows the results of an experiment where two quantities, x and y, were being measured.

x	1	3	3	4	5
y	1·5	6·7	14	24	41

The graph for this relationship is shown below.

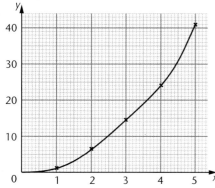

The graph looks rather like a parabola ($y = kx^2$) but, to confirm this, draw the graph of y against x^2.

The table for this is shown below.

x	1	2	3	4	5
x²	1	4	9	16	25
y	1·5	6·7	14	24	41

The graph for y against x^2 is shown below.

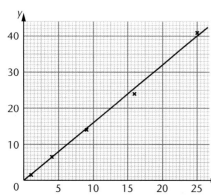

This graph is approximately a straight line and so the line of best fit can be drawn.

This line has a gradient of $\frac{40}{25} = 1·6$ and goes through the origin.

Since the graph is for y against x^2 the equation of a straight line is $y = mx^2 + c$ and in this case $m = 1·6$ and $c = 0$.

The equation that fits the relationship is $y = 1·6x^2$.

EXAMPLE 5

In an experiment the quantities x and y are measured. The table of values is given below.

x	1	2	5	10	15	20
y	6·2	4·3	3	2·9	2·6	2·5

a) Draw a graph of y against x.

b) It is thought that the relationship between y and x is of the form $y = \dfrac{a}{x} + b$

Confirm this by plotting a graph of y against $\dfrac{1}{x}$.

c) Draw a line of best fit and use it to estimate the values of a and b.

a)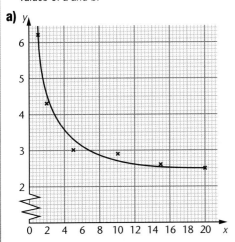

Since the graph is similar to the $y = \dfrac{1}{x}$ graph translated in the y direction, the given relationship seems likely.

Exam tip

Make a table of values of $f(x)$ and y before plotting. Make sure that your scale for $f(x)$ is linear. It is very easy to mark a scale off, e.g. as 1, 4, 9, 16, 25... instead of 5, 10, 15, 20, 25... When drawing the axes, allow for the fact that b could be negative by leaving room for negative numbers on the y-axis.

b) The table for the second graph is

x	1	2	5	10	15	20
$\dfrac{1}{x}$	1	0·5	0·2	0·1	0·067	0·05
y	6·2	4·3	3	2·9	2·6	2·5

This graph is drawn below.

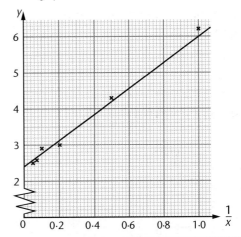

Since the graph is reasonably linear this confirms the relationship.

c) The gradient of the line of best fit
$= (6 - 2\cdot4) \div 1 = 3\cdot6$. So $a = 3\cdot6$.
The y intercept is 2·4 so $b = 2\cdot4$.

The relationship is therefore $y = \dfrac{3\cdot6}{x} + 2\cdot4$.

It must be emphasised that this result is only conjecture since the data is experimental, and that the values of a and b are only estimates.

In general, if a relationship is thought to be of the form $y = mf(x) + c$ this can be tested by plotting y against $f(x)$.

m and c can be estimated using the line of best fit.

m = the gradient and c = the y intercept.

135

Chapter 2 *Finding formulae that connect data*

1 State the variables you would need to plot to obtain a straight-line graph of the following equations.

a) $y = 3 \cdot 6x^2 + 5$ **b)** $V = 10d^3$

c) $C = 5\sqrt{m} + 3$

2 Use these graphs to find the equations connecting the variables.

a)

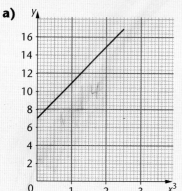

b)

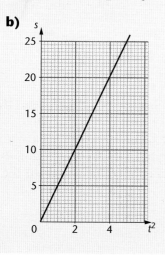

3 Find the equations of these graphs.

a)

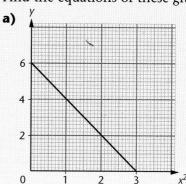

b)

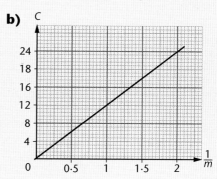

4 **a)** Complete the missing row in this table of values.

x	2	5	6	9
x²				
y	2	65	98	233

b) Plot a graph of y against x^2 and hence find the equation connecting y and x.

Exercise 2.2A cont'd

5

t	1	2	3	4	5
y	3·0	4·7	7·8	12·1	17·4

There is an approximate equation connecting y and t of the form $y = at^2 + b$. Plot the graph of y against t^2 and use your graph to estimate the values of a and b.

6 An equation of the form $x = at^3 + b$ connects the values in this table. Draw a suitable straight-line graph and obtain this equation.

t	1	1.5	2	2.5	3
x	6·4	7·35	9·2	12·25	16·8

7 One of the following experimental results has been recorded wrongly. The variables should be connected by an equation of the form $s = \dfrac{a}{x} + b$.

x	1	1.5	2	3	4	8
s	11.9	9.3	8.4	6.7	6.0	4.1

Plot suitable variables to obtain a straight line graph for the formula. Ring the point which is incorrect and draw a line of best fit through the remaining points. Hence find the equation connecting s and x.

8 A company installing garden ponds decides on a pricing structure of a fixed price, £P, plus a multiple of the square of the length, L m, of the pond.

Here are examples of some of their charges:

L m	1·2	1·5	2·0	2·5
£P	62·90	91·25	152·50	231·25

By drawing a suitable graph, or otherwise, find the equation connecting P and L.

9

d	0·26	0·58	0·86	0·93
m	26	293	954	1210

The diameters, d cm, and mass, M g, of some ball bearings were measured. Plot a graph of M against d^3, draw the line of best fit and hence find the relationship connecting M and d.

10 The power, P kilowatts, needed to achieve different speeds, v km/h, for a certain racing car are shown in the table.

v	120	140	160	180	200	220
p	50	54	58	62	65	68

The engineer thinks these are related by the formula $P = k\sqrt{v}$ for some value of k. Draw an appropriate graph to test this theory, stating the equation connecting P and v if the theory is verified.

EXERCISE 2.2B

1 State the variables you would need to plot to obtain a straight line graph of the following equations:

a) $s = 4t^3 + 6$

b) $A = \pi r^2$

c) $P = \dfrac{4}{x} + 3$

2 Use these graphs to find the equations connecting the variables.

a)

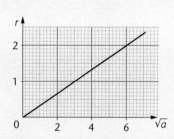

b)

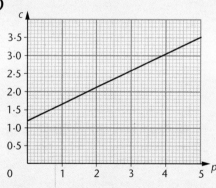

3 Find the equations of these graphs.

a)

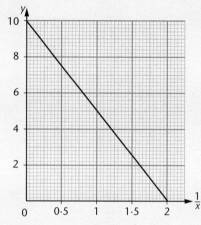

b)

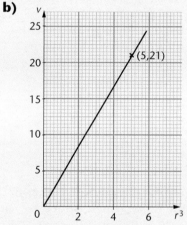

4 a) Complete the missing row in this table of values.

x	1	2	4	5
$\dfrac{1}{x}$				
y	6	4	3	2·8

b) Plot a graph of y against $\dfrac{1}{x}$ and hence find the equation connecting y and x.

Exercise 2.2B cont'd

5

x	1	2	3	4	5
C	⁻3	1·5	9	19·5	33

There is an equation connecting C and x of the form $C = ax^2 + b$.

Plot the graph of C against x^2 and use your graph to find the values of a and b.

6 By drawing a graph, or otherwise, calculate the values of a and b in the equation $y = ax^3 + b$ satisfied by the following values:

x	1	2	5	10
y	⁻37	⁻16	335	2960

7 The following values are obtained in an experiment concerning distance travelled (s m) and time taken (t seconds):

t	0·8	1·5	3·2	3·7
s	2·6	9·2	40·9	54·9

Plot the values of s against t^2 and draw the line of best fit. Hence obtain the equation connecting s and t.

8

t	1	2	3	4	5
s	1·0	1·25	1·44	1·6	1·74

a) How can you tell from the table that s and t are not linearly related?

b) It is thought that s is proportional to $\sqrt{t}$. Draw an appropriate graph and obtain the equation connecting s and t. Explain, giving your reason, whether s is proportional to $\sqrt{t}$.

9 Some rope is sold in different sizes. These sizes are the circumference, c millimetres, of the rope measured when it is unstretched. This table shows the price, p pence, per metre for the different sizes.

c	10	20	30	40	50
p	24	36	56	84	120

a) Plot a graph of p against c^2 and hence obtain the equation connecting p and c.

b) What is the cost per metre of one of these ropes with circumference 70 mm?

Key ideas

- When writing general rules in algebraic form, you must:
  a) define all the variables precisely, including the units
  b) make sure that the correct rules of algebra are followed in writing the formula that fits the rule.
- If a relationship is thought to be of the form $y = m\,f(x) + c$, this can be tested by plotting $f(x)$ against y.

  m and c can be estimated using the line of best fit.

  m = the gradient and c = the y intercept.

139

Chapter 2 *Finding formulae that connect data*

3 Surveys and sampling

You should already know

- **how to analyse data, e.g. calculate the mean, median and mode and range**
- **how to design questionnaires**
- **how to carry out pilot surveys**
- **how to display information and results.**

As explained in the Introduction, as part of your GCSE assessment you will be expected to undertake an extended piece of statistical work. The criteria that will be used to assess this are also given, in a simplified form, in the Introduction. However, important aspects of the criteria are referred to throughout this chapter.

As an overview it might be helpful to consider the four aspects of the data handling cycle.

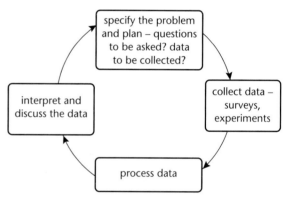

It is important to make sure that, at the end of your work, the conclusions you make relate back to the initial aims. In other words, 'Have you answered the question?' and that you try to present a well-structured and coherent report.

ACTIVITY 1

- Think back to a statistics project you have done in the past.
  - What went well?
  - What was difficult?
  - Was there anything you wished afterwards that you had done differently?
  - How can you apply what you learned then to your next project?
- See how this fits into the data handling cycle.

Surveys

A survey can be thought of as an investigation to establish something, e.g.:

- people's preferences, likes/dislikes or beliefs
- the number of plants or animals in a particular location
- the presence of oil or gas (a commercial example).

Remember that you should explain how you are going to collect data, consider any practical problems in carrying out the survey, and explain how you will plan for these.

You should also use, and explain why you are using, a pilot survey or a pre-test of a questionnaire, etc.

Questionnaire design

One common method of obtaining information from people is to use a questionnaire. Questionnaires need to be designed to provide answers that are easily analysed.

- Those with 'yes' or 'no' answers are clearly easy to analyse but remember that you must try to avoid writing questions which have an 'or' statement in them. For example if you were doing a survey about the number of people who were vegetarian, asking a questions such as 'do you eat meat or vegetables' will not provide any useful answers!

- One way is to ask questions giving options (boxes) from which the subject selects a choice. This allows the information to be counted and analysed. Sometimes a box can be given as 'anything else' and a line left for comments, but it is a good idea to avoid this.

- Avoid questions which
  a) are not to the point (i.e. about the subject)
  b) are embarrassing or biased.

EXERCISE 3.1

1 Julie is designing a questionnaire for a survey about her local supermarket.

She has decided to test two hypotheses:

a) local people visit the supermarket more often than people from further away

b) local people spend less money per visit.

Write two questions which could help her test each hypothesis.

Each question should include at least three options for a response. The people interviewed are being asked to choose one of these options.

2 A Sports and Leisure Club offers a range of facilities for many different sporting activities. As part of their reviewing systems the committee issue questionnaires to a sample of members.

One question said: 'If you are a regular tennis player and think better facilities are needed, how many extra courts should be provided and what other improvements would you like to see?'

a) Write down at least two faults in this question.

b) How would you improve the question? Write down your improved question.

c) How should the committee select a representative sample?

Sampling

This section provides an introduction to sampling and the methods used to obtain a sample.

In data handling the word '**population**' is used for a collection, set or group of objects being studied.

A '**sample**' is a smaller group (a subset), selected from the population. If the population is large it is not usually possible or practicable to collect data on every member of that population so one or more samples will be surveyed and conclusions will be drawn from these which will then be applied to the whole population.

In your work you should show that you are using a sample of an adequate size.

If the structure or composition of the population is known then it is important to ensure that the sample (or samples) represents that population and thus any variations in that population should be reflected in that sample – which is therefore called a **representative sample**.

You must explain why you have chosen a particular sampling method, why the sample is of the size that it is, and also what effects the nature of the sample may have on your findings.

There are various methods of choosing a representative sample.

1 Systematic sampling

An example of this method would be the selection of a 10% sample by going through the population picking every tenth item or individual. The drawback is that this would only provide a representative sample if the population was arranged in a random way and not in a way that might introduce bias. (Bias is described later.)

2 Attribute sampling

In this method the selection of the sample is made by choosing some attribute which is totally unrelated to the variable being investigated. Choosing a sample to investigate any relationship between head size and height from a list of people on the basis of their birthday being the first of the month would be an example of this.

3 Stratified or quota sampling

The population is divided into strata or subgroups and the sample chosen to reflect the properties of these subgroups. For example, if the population contained three times as many people under 25 as over 25 then the sample should also contain three times as many people under 25. The sample should also be large enough for the results to be significant.

4 Random sampling

However, if there is no information about the characteristics of the population – e.g. no knowledge about the ages and sex of the people in the population, or about the colours and sizes of the objects in a population, then a sample must be selected on the basis that all items are equally likely to be chosen. This is called random sampling. To ensure a sample is random and as accurate as possible, ideally the sampling should always be repeated a number of times and the results averaged.

You will find examples of 'sampling' taking place throughout the year:

- in politics, with opinion polls reporting on the popularity of the political parties, especially in the weeks before local and general elections
- in market research – 'eight out of ten owners said their cats preferred Whiskas', or whether the building of the Millennium Dome was a good use of money, etc.

You may well have seen market researchers interviewing people in the street.

ACTIVITY 2

- Treat the teaching class as a 'population' and calculate the average height exactly.
- Use each of methods 1 to 4 to identify a sample of about 6 to 8 students and see how close their average height is to the class average.

EXERCISE 3.2

1 A soap powder manufacturer wants to know what percentage of the population use its washing powder.

Would it be likely to obtain a representative sample by asking:

a) people leaving a public house at 10 p.m. on a Friday night

b) people leaving a supermarket at 7·30 p.m.

c) people leaving a supermarket between 9 a.m. and noon each day for a week

d) people getting off a commuter train on their way home from work?

Give reasons for each of your answers.

Can you suggest a way of obtaining a representative sample?

2 You might be able to try the following experiment.

a) Put 100 counters in a bag. The counters should be the same size but of differing colours e.g. some red, some blue etc.

b) Chose a colour (e.g. red).

c) Select 10 counters without looking at them and make a note of the number of counters selected which are of your chosen colour (e.g. 3 red ones). Return the counters to the bag.

d) Multiply by 10 to predict the number of red counters in the bag ($10 \times 3 = 30$). (Write this number down so you don't forget it.)

e) Repeat steps **c)** and **d)** another 9 times giving a total of 10 experiments (e.g. 20, 30, 40, 40, 50, 20, 30, 40, 40).

f) Find the mean of the 10 experiments (e.g. $\frac{340}{10} = 34$).

This figure ought to be close to the actual number of red counters in the bag.

How does the answer you obtain compare with the actual number of counters of your chosen colour in the bag?

3 All schools could be improved. Write a questionnaire to give to a sample of pupils in your year or the whole school. Try to write questions which can be analysed easily. How are you going to select the sample?

EXERCISE 3.3

1 Comment on the following ideas for obtaining a random sample.

Give a better method if you can.

a) A random sample of all the adults in a town is required.

Method: stand outside a supermarket and stop every tenth person who leaves between the hours of 9 a.m. and 3 p.m.

b) A random sample of the students at your school is required.

Method: ask all the students to 'sign up' if they want to take part in a study and promise to pay £1 to all who are chosen. Choose at random until the required number of students is obtained.

2 Which of the following would you study by sampling?

a) The average life of a torch battery.

b) The top ten singles for, say, last week.

3 Which sampling method would you use in the following situations? Describe how to select a suitable sample in each case.

a) What proportion of the constituency you live in will vote for a particular party?

b) What is the average height of a student in your year group or school? Does the height of girls differ much from the height of boys?

c) What is the number of trees in a local wood?

d) What is the number of students in your school who would go to a school disco? Or a fair?

e) How much homework do students do? Does it increase as you get older?

Bias

If each 'item' in the population does not have an equal chance of being selected for the sample, then the sample is said to be **biased**.

Examples of biased samples could be:

1 studying illness in the elderly by obtaining information from a few residential homes (this is likely to be biased because **a)** it may not be representative of the population – only those who can afford to or want to might live in such homes – and **b)** infectious illnesses are more likely to be spread when people live close together)

2 investigating the pattern of absence from a school by studying the registers in December (might be biased because children are more likely to be ill in the winter months compared with, say, the summer months; older students might be absent for interviews and the pattern of any truancy might vary)

3 finding options about school dinners by asking the first 50 students in the dinner queue one day (might be biased because only those eating school dinners might be asked – students might be queuing by year groups and so not be representative).

Your work should show what measures you took to avoid any bias in the sample you have chosen.

Stratified random sampling

As mentioned earlier, sampling which is representative of the whole population is called stratified sampling. The method used is as follows:

• Separate the population into appropriate categories or strata, e.g. by age.

• Find out what proportion of the population is in each stratum.

• This can be done by random sampling and so the technique is known as stratified random sampling.

EXAMPLE 1

The 240 students in Year 9 of a school are split into four groups for games. 90 play cricket, 70 play tennis, 30 choose athletics and the remaining 50 opt for volleyball.

Use a stratified random sample of 40 students to estimate the mean weight of all 240 students.

The sample size from each of the four groups must be in proportion to the stratum size, so the 40 students are selected as follows:

Cricket $\quad \dfrac{90}{240} \times 40 = 15$

Tennis $\quad \dfrac{70}{240} \times 40 = 11 \cdot 67$ i.e. 12

Athletics $\quad \dfrac{30}{240} \times 40 = 5$

Volleyball $\dfrac{50}{240} \times 40 = 8 \cdot 33$ i.e. 8

Within each sample the actual students will be selected randomly (random sampling is discussed later).

The mean weights for each sample are found to be:

| Cricket | 54·6 kg | Tennis | 49·7 kg |
| Athletics | 53·1 kg | Volleyball | 47·9 kg |

so the mean weight for all 240 students =

$$\dfrac{54 \cdot 6 \times 15 + 49 \cdot 7 \times 12 + 53 \cdot 1 \times 5 + 47 \cdot 9 \times 8}{40}$$

$$= 51 \cdot 6 \, \text{kg}$$

This will be an estimate for the mean weight of the population of 240 students.

EXAMPLE 2

A different form of sampling is illustrated by this example.

The natterjack toad is an increasingly threatened species of toad.

Scientists want to find out how many of these toads live in and around a pond.

To do this they catch 20 and mark them in a harmless way. The toads are then released.

Next day another 20 are caught: 5 of these toads have already been marked, in other words a sample of 25% (5 out of a sample of 20) are marked. But 20 toads were marked initially.

This suggests that 25% of the population is about 20.

$$\dfrac{25}{100} \times P = 20 \text{ therefore } P = 80$$

Therefore the total population is 80.

EXERCISE 3.4

1 Amy wants to investigate the spending habits of students at her school. The number of students in each year group is as follows:

Year group	7	8	9	10	11
Number of students	208	193	197	190	184

Explain how Amy obtains a stratified sample of 100 students for her survey.

2 The table shows the number of boys and girls in Year 10 and Year 11 of a school.

	Year 10	Year 11
Boys	120	134
Girls	110	100

The headteacher wants to find out their views about changes to the school uniform and takes a stratified random sample of 50 pupils from Year 10 and Year 11.

Calculate the number of pupils to be sampled from Year 11.

3 Scientists need to estimate the number of fish in Lake Hodder. They catch and 'tag' 450 fish and then release them back into the lake.

Over the next few days and at various locations they catch samples and count the number of fish that are tagged.

Day	Sample size	Number 'tagged'
1	36	6
2	38	6
3	40	8
4	32	6

Use these values to estimate the total number of fish in the lake.

4 Repeat Exercise 3.2 question 2 (knowing the number of, say, red counters). Take four samples and use the number of red counters in each sample to estimate the total number of counters.

How close is this estimate to the known answer of 100?

Random numbers

Various methods are available in order to select the items for a random sample. These include:

- the random number facility included on scientific and graphical calculators – see your own calculator manual to find out how to use it to generate random numbers
- using random number tables.

This is part of a random number table.

215656423238732265665625189987922132260659546008487403306678056086556696
665100454587964544448845554485040345648456548978554416451232358885689857
884653316596662365752639526892332546154812153215546555333154881841638961
568664569569665548416963166156435615368663156646345765989434879315693631
663366631666645999696399973393631596406556862589625548963225879452132595
237412365358625562561354598753314633221489625658966654896524488854475221
132335523232545623256654879577522963221486322156332152323255562322156255
48563258984436544326542136545236

The following example illustrates the method.

You might like to try this if you have time, either on your own or with friends. However, do work through the example so that you will understand the method even if you haven't time to do all the calculations.

EXAMPLE 3

Plant Laboratories have produced a new variety of a flowering plant and are interested in the size of the flower heads. The diameters of the flowers on 50 plants are measured and shown as circles on the diagram below. The circles are numbered for reference.

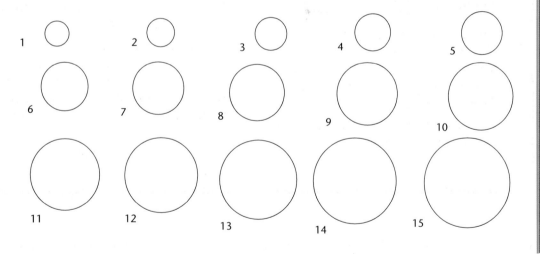

Example 3 cont'd

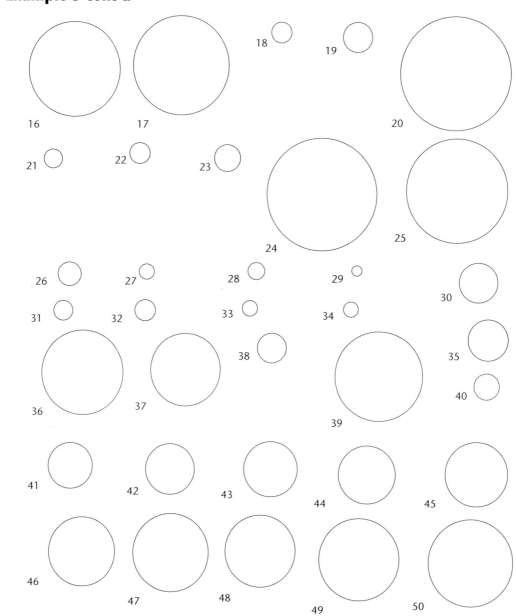

Example 3 cont'd

1 Select a sample of five circles that you think are representative. Measure their diameters and calculate the mean.

2 Repeat for another two samples and find the overall mean.

(Collect the results from any others in your class who are also doing this activity.)

3 Using a random number table, such as the one above,

a) choose a starting number in this table. Once you have fixed this you must move along the row or down the column and not 'jump about'. Divide the numbers into pairs, starting with your chosen number, to represent the two-digit reference numbers of the flowers, so 03 would represent picture 3. Discard or ignore any two-digit numbers that are greater than 50.

b) write down the first five two-digit numbers that you get, measure the diameters of those circles and calculate the mean diameter.

or use the random number generator on your calculator and follow steps a) and b).

The actual mean diameter for the 50 flowers is 1·43 cm.

(i) how close were the means calculated by both methods?

(ii) how did you judge 'representative'. In other words what criteria did you use?

Discuss this with others in your class who have completed the experiment.

EXERCISE 3.5

1 The table shows the times taken by 100 pupils to solve a mathematics problem. The times are all in seconds.

14	20	73	35	28	39	25	17	16	23
20	7	13	30	39	36	17	35	57	26
150	39	25	27	25	40	39	62	47	25
22	16	32	46	29	21	57	10	122	81
90	34	12	68	28	81	32	47	35	37
39	40	23	46	25	43	74	53	24	51
12	30	93	26	17	21	32	37	33	42
93	40	18	55	11	56	34	67	13	15
104	21	25	49	35	18	15	47	26	57
38	92	59	12	32	46	36	25	71	35

Using a sample, estimate the mean and median time taken to solve the problem.

a) What sample size would you use and why?

b) Use your sample to obtain the estimates of the mean and median times.

Exercise 3.5 cont'd

2   Here is a list of projects that could be attempted or at least planned.

In each case you will need to decide:

• who or what to sample, i.e. how the sample is to be found.

• what questions need asking or parameters need measuring.

Remember to analyse the results, explaining why you chose to calculate the median, mean or mode; why you chose to present the results in any particular way, etc.

a) What is an average student?

b) Old people are more superstitious than young people.

c) More babies are born in winter than in summer.

d) Do tall people weigh more than short people?

e) Estimate the number of blades of grass on, e.g. the school playing field, a football pitch, or your lawn at home.

f) Any ideas suggested by any of the questions in this chapter.

Key ideas

You should be able to:

○   design and write questionnaires

○   decide an appropriate sample size and how to select this sample.

You should understand:

○   random sampling and the use of random number tables or random number generators

○   stratified or quota sampling.

Rational and irrational numbers

You should already know

- what natural numbers and integers are
- what terminating and recurring decimals are
- the 'dot' notation for recurring decimals
- how to multiply out and simplify brackets like $(a + b)(c + d)$.

Rational numbers

To solve equations such as $x + 3 = 8$, the only set of numbers necessary is the set of **natural numbers** 1, 2, 3, 4, 5, ...

To solve equations such as $x + 8 = 3$, it is necessary to introduce negative numbers to our set of numbers.

This gives the **integers** ... $^-4$, $^-3$, $^-2$, $^-1$, 0, 1, 2, 3, 4, ...

To solve equations such as $4x = 3$ or $3x = 8$ or $^-7x = 5$, numbers like $\frac{3}{4}$, $\frac{8}{3} = 2\frac{2}{3}$, $\frac{^-5}{7}$ ... need to be introduced.

These numbers are called **rational numbers**. A rational number is one that can be written as a fraction with integers as numerator and denominator.

As well as the obvious fractions, rational numbers include:

natural numbers, e.g. 5 can be written $\frac{5}{1}$

integers, e.g. $^-6$ can be written $\frac{^-6}{1}$

terminating decimals, e.g. 3·24 can be written as $\frac{324}{100}$

recurring decimals, e.g. 0·6 = 0·666 66 ... can be written as $\frac{2}{3}$.

Irrational numbers

The question arises as to what numbers are left?

To solve equations like $x^2 = 2$, numbers such as $\sqrt{2}$ need to be introduced.

$\sqrt{2} = 1·414\,213\,562 ...$

This is a decimal that neither terminates nor recurs, so it cannot be written as a fraction.

These numbers are called **irrational numbers** as they cannot be written as a fraction with two integers. They are decimals that go on forever without recurring.

Irrational numbers include:

a) all non-exact square roots, e.g. $\sqrt{7}$ but NOT $\sqrt{16}$ or $\sqrt{\frac{9}{16}}$

b) special numbers that occur in mathematics. The only one that you have met so far is 'π', which is 3·141 592 654 ... . Another that is on your calculator but which you will not use unless you study mathematics beyond GCSE is 'e' which is 2·718 281 828 ... .

It can be proved that e.g. $\sqrt{2}$ is irrational but that is not necessary at this stage.

Since the rational numbers include the natural numbers and integers, and the irrational numbers are all those numbers that are not rational, these two sets comprise all the real numbers that you will need to use at GCSE.

ACTIVITY 1

Find a proof that $\sqrt{2}$ is irrational or try to prove it yourself.

> There is an elegant proof by contradiction which starts:
>
> Assume $\sqrt{2}$ is rational. Then $\sqrt{2}$ is $\frac{a}{b}$ for some integers a and b whose only common factor is 1.

Squaring: $2 = \frac{a^2}{b^2} \ldots$

Hint

If a number is expressed as $2n$, what do you know about that number?

EXAMPLE 1

State which of these numbers are irrational.

a) $\frac{3}{4}$ **b)** $\frac{2}{3}$ **c)** $\sqrt{11}$ **d)** 2π **e)** 3·142   **f)** $^-1\frac{1}{4}$.

Clearly **a)** and **b)** are rational.

> 3·142, although an approximation to π, is a terminating decimal and can be written as $\frac{3142}{1000}$.
> $^-1\frac{1}{4}$ can be written as $\frac{^-5}{4}$

The only two of these that cannot be written as fractions are **c)** and **d)**, so these are irrational.

Recurring decimals and fractions

Fractions can easily be converted to decimals by dividing the numerator by the denominator.

e.g. $\frac{5}{9} = 5 \div 9 = 0.555\,555\ldots = 0.\dot{5}$.

$\frac{256}{111} = 256 \div 111 = 2.306\,306\,306\ldots = 2.\dot{3}0\dot{6}$.

Converting recurring decimals to fractions is somewhat more difficult. It may be worth remembering that, as above, $0.\dot{5} = \frac{5}{9}$ and similarly $0.\dot{7} = \frac{7}{9}$, etc., but if more figures recur a more formal method is needed.

This method is illustrated in the following example.

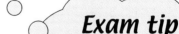

Exam tip

The dots over the digits show how many recur. If it is more than two, only the first and last has a dot.

EXAMPLE 2

Express $0.\dot{4}\dot{2}$ as a fraction in its lowest terms.

Let $r = 0.\dot{4}\dot{2}$	$= 0.42424242 \ldots$
Multiply by 100	$100r = 42.42424242\ldots$
Subtract	$\underline{r = 0.42424242\ldots}$

$$99r = 42$$

$$r = \frac{42}{99} = \frac{14}{33}$$

So $0.\dot{4}\dot{2} = \frac{14}{33}$.

ACTIVITY 2

Investigate patterns of recurring decimals.

- In groups, without a calculator, share the work to find as decimals: $\frac{1}{7}, \frac{2}{7}, \frac{3}{7}, \frac{4}{7}, \frac{5}{7}, \frac{6}{7}$, and compare your results.
- Similarly, have a quick look at what happens with $\frac{1}{9}, \frac{2}{9}$, etc., and $\frac{1}{11}, \frac{2}{11}$ etc.
- More difficult, but more interesting, are the thirteenths and seventeenths. A calculator may be of limited help here.

The method always works provided you multiply by the correct power of 10.

If one figure recurs multiply by 10.

If two figures recur multiply by 100.

If three figures recur multiply by 1000 and so on.

EXAMPLE 3

Write $0.4\dot{2}0\dot{7}$ as a fraction in its lowest terms.

Let $r = 0.4\dot{2}0\dot{7}$ $\quad = \quad 0.4207207207 \ldots$ Note: three figures recur.

Multiply by 1000 $\quad 1000r = 420.7207207207\ldots$

$\qquad\qquad\qquad r = \quad 0.4207207207\ldots$

Subtract $\qquad\quad 999r = 420.3$

$$r = \frac{420.3}{999} = \frac{4203}{9990} = \frac{1401}{3330} = \frac{467}{1110}$$

So $0.4\dot{2}0\dot{7} = \frac{467}{1110}$

For numbers such as $3.4\dot{2}0\dot{7}$ the three can be added at the end.

So $3.4\dot{2}0\dot{7} = 3\frac{467}{1110}$.

EXERCISE 4.1A

1   State whether each of these numbers is rational or irrational, showing how you decide.

   a) $\frac{17}{20}$ **b)** $0\cdot46$ **c)** $\sqrt{\frac{4}{25}}$ **d)** $\sqrt{169}$ **e)** 5π

   f) $3\cdot14159$ **g)** $^-0\cdot2\dot{3}\dot{4}$ **h)** $5 + \sqrt{3}$ **i)** $^-6\sqrt{2}$ **j)** $\sqrt{\frac{4}{25}}$

2   Convert these fractions to recurring decimals using the dot notation.

   a) $\frac{7}{11}$ **b)** $\frac{3}{7}$ **c)** $\frac{3}{70}$ **d)** $\frac{23}{90}$ **e)** $\frac{2079}{4995}$

3   Convert these recurring decimals to fractions or mixed numbers in their lowest terms.

   a) $0\cdot\dot{2}$ **b)** $0\cdot4\dot{8}$ **c)** $0\cdot2\dot{3}$ **d)** $0\cdot1\dot{3}\dot{2}$ **e)** $2\cdot1\dot{8}$

EXERCISE 4.1B

1   State whether each of these numbers is rational or irrational, showing how you decide.

   a) $0\cdot49$ **b)** $0\cdot5\dot{3}$ **c)** $\sqrt{324}$ **d)** $\sqrt{27}$ **e)** $5\pi + 2$

   f) $^-2\cdot718$ **g)** $\frac{4\pi}{3\pi}$ **h)** $2\sqrt{3} + \sqrt{5}$ **i)** $\sqrt{2} - 7$ **j)** $\sqrt{1\frac{7}{9}}$

2   Convert these fractions to recurring decimals using the dot notation.

   a) $\frac{7}{33}$ **b)** $\frac{5}{13}$ **c)** $\frac{5}{1300}$ **d)** $\frac{17}{36}$ **e)** $\frac{481}{1100}$

3   Convert these recurring decimals to fractions or mixed numbers in their lowest terms.

   a) $0\cdot\dot{7}$ **b)** $0\cdot4\dot{3}$ **c)** $0\cdot40\dot{2}$ **d)** $0\cdot23\dot{6}$ **e)** $0\cdot\dot{1}23\dot{4}$

Simplifying surds

An irrational number like $\sqrt{3}$ or $6 + 2\sqrt{5}$ is called a surd.

Surds can often be simplified by using the result

$\sqrt{(a \times b)} = \sqrt{a} \times \sqrt{b}$.

This result can be demonstrated using $\sqrt{36}$.

$\sqrt{36} = 6 = 2 \times 3 = \sqrt{4} \times \sqrt{9}$ so $\sqrt{36} = \sqrt{(4 \times 9)} = \sqrt{4} \times \sqrt{9}$.

EXAMPLE 4

Simplify $\sqrt{50}$.
$\sqrt{50} = \sqrt{(25 \times 2)} = \sqrt{25} \times \sqrt{2} = 5\sqrt{2}$

EXAMPLE 5

Simplify $\sqrt{72}$.
9 is a factor of 72 so $\sqrt{72} = \sqrt{9} \times \sqrt{8} = 3\sqrt{8}$,
but 4 is a factor of 8 so $3 \times \sqrt{8} = 3 \times \sqrt{4} \times \sqrt{2} = 3 \times 2 \times \sqrt{2} = 6\sqrt{2}$
Or, if you spot straight away that 36 is a factor of 72,
$\sqrt{72} = \sqrt{(36 \times 2)} = \sqrt{36} \times \sqrt{2} = 6\sqrt{2}$

> **Exam tip**
>
> Look for as large a factor of the number as possible which has an exact square root. In Examples 4 and 5, this is 25 and 36 respectively.

EXAMPLE 6

Simplify $\sqrt{12} \times \sqrt{27}$
Method 1 $\sqrt{12} = \sqrt{(4 \times 3)} = \sqrt{4} \times \sqrt{3} = 2 \times \sqrt{3}$
$\qquad\qquad \sqrt{27} = \sqrt{(9 \times 3)} = \sqrt{9} \times \sqrt{3} = 3 \times \sqrt{3}$
So $\sqrt{12} \times \sqrt{27} = 2 \times \sqrt{3} \times 3 \times \sqrt{3} = 2 \times 3 \times \sqrt{3} \times \sqrt{3}$
$\qquad\qquad\qquad\quad = 6 \times 3 = 18$
Method 2 $\sqrt{12} \times \sqrt{27} = \sqrt{(12 \times 27)} = \sqrt{324} = 18$

Although Method 2 is probably easier if you have a calculator, Method 1 is probably easier if the question comes on the non-calculator paper.

Note: by definition of what we mean by a square root, $\sqrt{a} \times \sqrt{a} = a$.

This question also illustrates the fact that the product of two irrational numbers can be rational.

Manipulation of numbers like $a + b\sqrt{c}$

A number like $2 + \sqrt{3}$, which is the sum of a rational number and an irrational number, is irrational.

This is because $2 + 1{\cdot}732\,050\,808... = 3{\cdot}732\,050\,808$ which is itself a decimal which goes on forever without recurring and so is irrational.

These numbers can be manipulated using the ordinary rules of algebra and arithmetic.

EXAMPLE 7

If $x = 5 + \sqrt{3}$ and $y = 3 - 2\sqrt{3}$, simplify:

a) $x + y$ **b)** $x - y$ **c)** xy

a) $x + y = 5 + \sqrt{3} + 3 - 2\sqrt{3} = 5 + 3 + \sqrt{3} - 2\sqrt{3} = 8 - \sqrt{3}$

b) $x - y = 5 + \sqrt{3} - (3 - 2\sqrt{3}) = 5 + \sqrt{3} - 3 + 2\sqrt{3}$

$ = 5 - 3 + \sqrt{3} + 2\sqrt{3} = 2 + 3\sqrt{3}.$

These two results illustrate the fact that when adding and subtracting these numbers, you can deal with the rational and irrational parts separately.

c) $xy = (5 + \sqrt{3})(3 - 2\sqrt{3}) = 15 - 10\sqrt{3} + 3\sqrt{3} - 2 \times \sqrt{3} \times \sqrt{3}$

$ = 15 - 2 \times 3 - 10\sqrt{3} + 3\sqrt{3}$

$ = 9 - 7\sqrt{3}$

EXAMPLE 8

If $x = 5 + \sqrt{2}$ and $y = 3 - \sqrt{2}$, simplify:

a) $x + y$ **b)** y^2

a) $x + y = 5 + \sqrt{2} + 3 - \sqrt{2} = 5 + 3 + \sqrt{2} - \sqrt{2} = 8.$

Note that this result indicates that it is possible for the sum of two irrational numbers to be rational.

b) $y^2 = (3 - \sqrt{2})^2 = (3 - \sqrt{2})(3 - \sqrt{2})$

$ = 9 - 3\sqrt{2} - 3\sqrt{2} + \sqrt{2} \times \sqrt{2}$

$ = 9 + 2 - 6\sqrt{2}$

$ = 11 - 6\sqrt{2}$

Note that this result is also an application of the algebraic result

$(a + b)^2 = a^2 + 2ab + b^2.$

Rationalising denominators

When dealing with fractions it is often preferable to have the numerator as an irrational number rather than the denominator. This is particularly so when no calculator is available as it is far easier to divide a long decimal by a whole number than to divide a whole number by a long decimal.

It is possible, using the rules of fractions, to convert numbers with irrational denominators to ones with irrational numerators. This method is illustrated in the following two examples.

EXAMPLE 9

Rationalise the denominator in these irrational fractions.

a) $\dfrac{5}{\sqrt{2}}$ **b)** $\dfrac{7}{\sqrt{12}}$

a) Multiply the numerator and the denominator by $\sqrt{2}$. By the rules of fractions, since we have multiplied both the numerator and denominator by the same quantity, we have not changed the value of the number.

This gives $\dfrac{5 \times \sqrt{2}}{\sqrt{2} \times \sqrt{2}} = \dfrac{5\sqrt{2}}{2}$ and the denominator is now a rational number.

b) First simplify the denominator and then repeat the process in part **a)**, this time multiplying by $\sqrt{3}$.

$\dfrac{7}{\sqrt{12}} = \dfrac{7}{2\sqrt{3}} = \dfrac{7 \times \sqrt{3}}{2\sqrt{3} \times \sqrt{3}} = \dfrac{7\sqrt{3}}{2 \times 3} = \dfrac{7\sqrt{3}}{6}$

EXERCISE 4.2A

1 Simplify the following, stating whether the result is rational or irrational.

   a) $\sqrt{12}$ **b)** $\sqrt{1000}$ **c)** $\sqrt{45}$ **d)** $\sqrt{300}$ **e)** $\sqrt{75}$

   f) $\sqrt{8} \times \sqrt{2}$ **g)** $\sqrt{20} \times \sqrt{18}$ **h)** $\sqrt{20} \div \sqrt{5}$ **i)** $\sqrt{80} \times \sqrt{50}$ **j)** $\sqrt{75} \times \sqrt{15}$

2 If $x = 4 + \sqrt{3}$ and $y = 4 - \sqrt{3}$ simplify:

   a) $x + y$ **b)** $x - y$ **c)** xy.

3 If $x = 3 + \sqrt{5}$ and $y = 4 - 3\sqrt{5}$ simplify:

   a) $x + y$ **b)** $x - y$ **c)** x^2.

4 If $x = 5 + 2\sqrt{3}$ and $y = 4 - 3\sqrt{2}$ simplify:

   a) $x\sqrt{3}$ **b)** x^2 **c)** y^2.

5 Rationalise the denominator in these irrational fractions.

   a) $\dfrac{1}{\sqrt{2}}$ **b)** $\dfrac{2}{\sqrt{5}}$ **c)** $\dfrac{5}{\sqrt{7}}$ **d)** $\dfrac{11}{\sqrt{18}}$ **e)** $\dfrac{9}{\sqrt{20}}$.

6 Simplify the following including rationalising the denominator.

   a) $\dfrac{6}{\sqrt{8}}$ **b)** $\dfrac{6}{\sqrt{300}}$ **c)** $\dfrac{12}{\sqrt{75}}$ **d)** $\dfrac{\sqrt{48}}{\sqrt{18}}$.

7 Find an exact expression for the shaded areas between these two squares, simplifying as much as possible.

$\sqrt{15}$

$\sqrt{5}$

8 Find, as simply as possible, an exact expression for the area of this circle.

$\sqrt{7}$

EXERCISE 4.2B

1   Simplify the following, stating whether the result is rational or irrational.

   a) $\sqrt{40}$ **b)** $\sqrt{54}$ **c)** $\sqrt{98}$ **d)** $\sqrt{800}$ **e)** $\sqrt{363}$

   f) $\sqrt{27} \times \sqrt{3}$ **g)** $\sqrt{250} \times \sqrt{40}$ **h)** $\sqrt{108} \div \sqrt{2}$ **i)** $\sqrt{90} \times \sqrt{20}$ **j)** $\dfrac{\sqrt{60} \times \sqrt{20}}{\sqrt{12}}$

2   If $x = 5 + \sqrt{7}$ and $y = 3 - \sqrt{7}$ simplify:

   a) $x + y$ **b)** $x - y$ **c)** xy.

3   If $x = 4 + \sqrt{11}$ and $y = 9 - 2\sqrt{11}$ simplify:

   a) $x + y$ **b)** $x - y$ **c)** x^2.

4   If $x = 6 - 2\sqrt{5}$ and $y = 3 - 5\sqrt{3}$ simplify:

   a) $x\sqrt{5}$ **b)** $y\sqrt{3}$ **c)** x^2 **d)** y^2.

5   Simplify $\sqrt{2}(5 + 3\sqrt{2})^2$.

6   Rationalise the denominator in these irrational fractions.

   a) $\dfrac{1}{\sqrt{7}}$ **b)** $\dfrac{3}{\sqrt{2}}$ **c)** $\dfrac{5}{\sqrt{11}}$ **d)** $\dfrac{7}{\sqrt{50}}$ **e)** $\dfrac{9}{\sqrt{32}}$

7   Simplify the following, including rationalising the denominator:

   a) $\dfrac{10}{\sqrt{5}}$ **b)** $\dfrac{15}{\sqrt{50}}$ **c)** $\dfrac{20}{\sqrt{32}}$ **d)** $\dfrac{12}{\sqrt{20}}$

8   Find the exact value of x, expressing your answer as simply as possible.

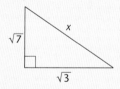

9   Find exact expressions for the total area of this shape formed from a square and a semicircle, simplifying as much as possible.

Key ideas

- A rational number is a number which can be written as a ratio or fraction with both numerator and denominator as integers.

- An irrational number is a number which cannot be written as a ratio or fraction with both numerator and denominator as integers. It is a number which, as a decimal, does not terminate or recur.

- Any recurring decimal can be written as a fraction.

- Surds can be simplified using $\sqrt{(a \times b)} = \sqrt{a} \times \sqrt{b}$

- Numbers which are the sum of a rational part and irrational part $(a + b\sqrt{c})$ can be dealt with using the normal rules of algebra.

- To rationalise a fraction with an irrational denominator, multiply the numerator and the denominator by the surd that is in the denominator

  e.g. $\dfrac{5}{2\sqrt{3}} = \dfrac{5}{2\sqrt{3}} \times \dfrac{\sqrt{3}}{\sqrt{3}} = \dfrac{5\sqrt{3}}{6}$.

A1 Revision exercise

1 The size, y, of a population of flies after t days was given by $y = 100 \times 1 \cdot 2^t$.

a) What was the size of the population at $t = 0$?

b) What was the size of the population after 5 days?

c) Use numerical methods to find the number of days it took for the population to reach 1000, assuming this rate of growth continued. Give your answer to 1 d.p.

2 The curve $y = ab^x$ passes through (0, 10) and (2, 6·4). Find the values of a and b.

3 In a chemical reaction, the mass of a chemical present is decreasing by 5% per minute. Initially there is 20 g of the chemical. Find, in minutes correct to 1 d.p., the time that passes before there is 2 g of this chemical left.

4 Here is a sequence

4, 6, 9, 13·5, 20·25, ...

Find:

a) the sixth term

b) the nth term

c) the 15th term correct to 3 d.p.

5 In the diagram there are 4 points on the circle and 6 lines joining them.

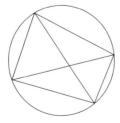

a) Find a formula for the number of lines joining n points on a circle.

b) Use the formula to find the number of lines joining 30 points on a circle.

6 Rushna has £P to invest in a bank which gives compound interest annually.

Find the formula for the amount the investment will be worth at the end of n years if the interest rate is

a) 4%   **b)** 6%   **c)** r%.

7 In an experiment the quantities x and y are measured. The table of values is given below.

x	5	10	15	20	25
y	0·7	5·5	16·3	43·5	87·2

It is thought that y is proportional to x^3.

a) Draw a graph to confirm this.

b) Draw a line of best fit and use it to find an equation connecting y and x.

8 The distance, s metres, travelled by a car in t seconds, is measured for several values of t.

From the results, a graph of s against t^2 is a straight line passing through (0, 0) and (4, 12). Find the equation connecting s and t.

9 Safia surveyed pupils in her school to find out their views about background music in shops.

The size of each year group in the school is shown in the table.

Year group	Boys	Girls
7	84	66
8	71	85
9	82	86
10	93	107
11	81	90
Total	**411**	**434**

Safia took a sample of 80 pupils.

a) Should she have sampled equal numbers of boys and girls in Year 7? Give a reason for your answer.

b) Calculate the number of pupils she should sample in Year 7.

10 A mobile phone company calls 200 people, chosen at random, who subscribe to their company, to find out how satisfied they are with the service they receive. Is this a satisfactory method of sampling? Give a reason for your answer.

11 Identify which type of sampling has been used in the following cases.

a) In order to determine whether the library facilities in a town are satisfactory, all the library cards are numbered and 100 questionnaires are sent out to the owners of cards selected, using random numbers.

b) A factory employs 1500 people on machines, 400 on packing and distribution and 300 in the offices. A sample is taken containing 15 machine operators, 4 people from packing and distribution and 3 office workers.

12 Sweet-tasting apples are used to make apple juice. An apple grower needs to find out how 'sweet' a crop of apples from an orchard is. He is advised to select a sample of 50 apples. The orchard consists of 1500 trees: each tree produces about 50 apples. The grower decides to pick 50 apples from one tree that he selects at random. Is this a satisfactory method? Give reasons for your answer.

13 State which of these numbers are rational and which are irrational.

a) $^-1 \cdot 6$ b) $0 \cdot \dot{7} \dot{3}$ c) $\frac{5\pi}{3}$
d) $7 + 2\sqrt{3}$ e) $1 \cdot 414$

14 Convert these fractions to recurring decimals using the dot notation.

a) $\frac{5}{11}$ b) $\frac{212}{999}$ c) $\frac{37}{495}$

15 Convert these recurring decimals to fractions or mixed numbers in their lowest terms.

a) $0 \cdot \dot{5} \dot{4}$ b) $3 \cdot 1 \dot{4} \dot{7}$ c) $0 \cdot 2 \dot{0} 3 \dot{4}$

Questions 16–20 should be done without a calculator.

16 Simplify:

a) $\sqrt{32}$ b) $\sqrt{150}$ c) $\sqrt{128}$
d) $\sqrt{12} \times \sqrt{75}$ e) $\sqrt{10} \times \sqrt{18}$
f) $\sqrt{72} \div 3$ g) $\sqrt{288} \times \sqrt{48}$.

17 If $x = 3 + \sqrt{7}$ and $y = 5 - 4\sqrt{7}$, simplify:

a) $x + y$ b) $x - y$ c) xy.

18 If $x = 5 + 2\sqrt{3}$ and $y = 5 - 2\sqrt{3}$ simplify:

a) x^2 b) y^2 c) xy.

19 Simplify $\sqrt{10}(5 + 2\sqrt{10})^2$.

20 Rationalise the denominator in the following, simplifying where possible.

a) $\frac{11}{\sqrt{2}}$ b) $\frac{15}{\sqrt{12}}$ c) $\frac{6}{\sqrt{27}}$.

5 Trigonometry in non-right-angled triangles

You should already know

● how to use Pythagoras' theorem and trigonometry in right-angled triangles.

All the trigonometry that you have learnt so far has been based on finding lengths and angles in right-angled triangles. However, triangles are often not right-angled. Some method needs to be established to find lengths and angles in these other triangles.

Notation

It is common practice in this work to use a single letter to represent each side and each angle of the triangle. We use a capital letter to signify an angle and a lower case letter for a side. It is usual for the side opposite an angle to take the same, lower case letter, as shown in the diagram.

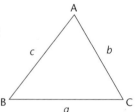

The rules

The two major rules for dealing with non-right-angled triangles are called the **sine rule** and the **cosine rule**. It will be obvious why they have these names when you see the formulae.

You do not have to memorise the formulae, as they are on the formulae sheet in the examination, but you have to know when and how to use them.

ACTIVITY 1

1 Find c, using right-angled trigonometry. It takes two steps, but is possible!

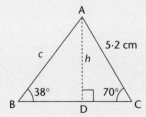

2 Now have a go at a general result, using the same processes as in part 1 of this activity.

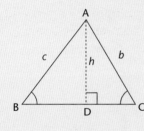

a) Use triangle ACD to find h in terms of b and C.

b) Use triangle ABD to find h in terms of c and B.

c) Equate your expressions for h and rearrange the result to show that

$$\frac{b}{\sin B} = \frac{c}{\sin C}$$

You have proved part of the sine rule!

The sine rule

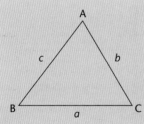

$$\frac{a}{\sin A} = \frac{b}{\sin B} = \frac{c}{\sin C}$$

$$or \quad \frac{\sin A}{a} = \frac{\sin B}{b} = \frac{\sin C}{c}$$

This formula is made up of *three* equal fractions. When using it you take two of the fractions. The two parts are chosen so that there is only one unknown value and three known. When finding a length the top formula is used; when finding an angle use the bottom one. This will mean that the unknown is on top of the fraction and will make the solution easier.

Chapter 5 *Trigonometry in non-right-angled triangles*

When to use the sine rule

When any two angles and one side are known.

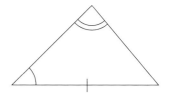

When two sides and the non-included angle are known.

Exam tip

Since you are finding a length, choose the formula with lengths on top. Choose pairs of angles and opposite sides where three of the four values are known and substitute into the formula.

EXAMPLE 1

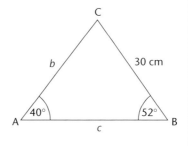

Find:

a) length b.

b) length c.

a) $\dfrac{b}{\sin 52°} = \dfrac{30}{\sin 40°}$

$b = \dfrac{30}{\sin 40°} \times \sin 52°$

$b = 36 \cdot 8$ cm

b) Before c can be found, angle C is needed.

$C = 180 - (40 + 52)$

$C = 88°$

$\dfrac{c}{\sin 88°} = \dfrac{3}{\sin 40°}$

$c = \dfrac{3}{\sin 40°} \times \sin 88°$

$c = 46 \cdot 6$ cm

Exam tip

Though you could use the pair b and B, you should always prefer to use values that are given rather than values that have been calculated.

EXAMPLE 2

Find angle C.

$$\frac{\sin C}{7\cdot1} = \frac{\sin 35°}{9}$$

$$\sin C = \frac{\sin 35°}{9} \times 7\cdot1$$

$$\sin C = 0\cdot4524...$$

$$C = \sin^{-1}(0\cdot4524...)$$

$$C = 26\cdot9°.$$

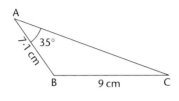

The third angle can now be found. Notice that it is obtuse (118·1°). You can find the third side using the sine rule again:

$$\frac{b}{\sin 118\cdot1} = \frac{9}{\sin 35}$$

Your calculator will find sin 118·1, so the problem may be solved. Further consideration of sines and cosines of angles greater than 90° will be given in Chapter 14.

EXERCISE 5.1A

1 Find c, A and a.

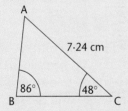

2 Find p, R and r.

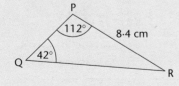

3 Find g, E and e.

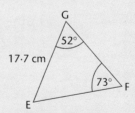

4 Find b, C and c.

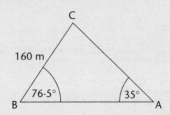

Chapter 5   *Trigonometry in non-right-angled triangles*

Exercise 5.1A cont'd

5 Find B, C and c.

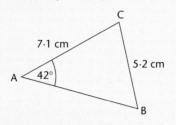

6 Find M, N and n.

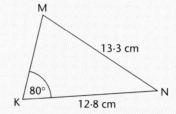

7 Find E, D and d.

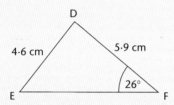

8 Calculate the largest angle of the triangle ABC given that A = 35°, a = 8·9 cm and c = 12 cm.

9

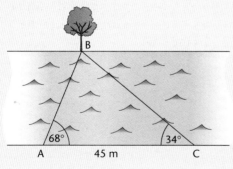

From two points, A and B, on horizontal ground, the angles of elevation of the top, T, of a vertical tower, TC, are 27° and 40° respectively. Given that AB = 30 m, find:

a) AT

b) BT

c) the height of the tower, TC.

10

A river has two parallel banks. Points A and C are on one side of the river, 45 m apart. The angles from these points to a tree, B, on the opposite bank are 68° and 34°, as shown in the diagram. Find:

a) AB

b) BC

c) the width of the river.

EXERCISE 5.1B

1 Find A, B and *b*.

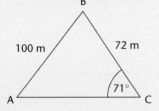

2 Find P, R and *r*.

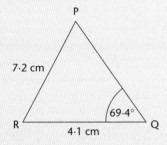

3 Find Y, Z and *z*.

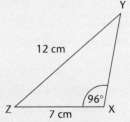

4 Find T, S and *s*.

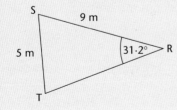

5 Find *b*, C and *c*.

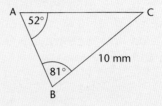

6 Find *y*, Z and *z*.

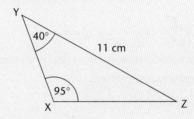

7 Find *s*, T and *t*.

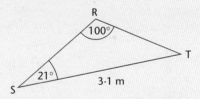

8 The top of a hill can be seen from A with an angle of elevation 40° and from C with an angle of elevation 55°. If AC is 120 m, calculate the distance AB.

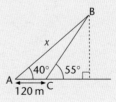

Exercise 5.1B cont'd

9 Town B is 45 km due East of town A. Town C is on a bearing 057° from town A and a bearing of 341° from town B. Find how far town C is from towns A and B.

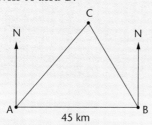

10 A child's slide, RST, is shown in the picture. Find the distance RT.

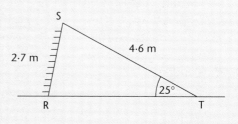

ACTIVITY 2

1 Another multi-step problem! Use the information in this triangle to work out *b*. This time you'll need Pythagoras' theorem as well as trigonometry – it is a three-step problem.

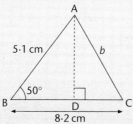

2 If you managed the challenge of generalising in part 2 of Activity 1, try this!

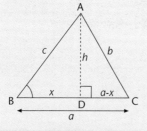

a) Use triangle ABD to find h^2 in terms of x and c.

b) Use triangle ACD to find h^2 in terms of b and $(a - x)$.

c) Equate your expressions for h^2 and make b^2 the subject. You should now have $b^2 = c^2 - x^2 + (a - x)^2$.

d) Work out the brackets and simplify your expression for b^2.

e) Use triangle ABD to find x in terms of B and c.

f) Substitute for x in your expression for b^2. You should have $b^2 = a^2 + c^2 - 2ac \cos B$.

You have proved the cosine rule!

The cosine rule

There are three versions of each of the formulae, but notice that they have exactly the same structure and pattern. Once again, there is one form to use when finding length and one for finding angles.

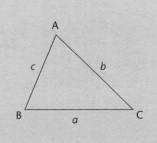

$$a^2 = b^2 + c^2 - 2bc \cos A$$
$$b^2 = c^2 + a^2 - 2ca \cos B$$
$$c^2 = a^2 + b^2 - 2ab \cos C$$

or

$$\cos A = \frac{b^2 + c^2 - a^2}{2bc}$$

$$\cos B = \frac{c^2 + a^2 - b^2}{2ca}$$

$$\cos C = \frac{a^2 + b^2 - c^2}{2ab}$$

When to use the cosine rule

When all three sides are known.

When two sides and the included angle are known.

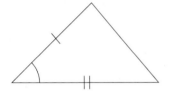

Exam tip

Notice that the cosine of an obtuse angle is negative. Your calculator will give cos 110 = ⁻0·342.... More detail is given in Chapter 14

EXAMPLE 3

Find c.

$c^2 = 4\cdot2^2 + 3\cdot7^2 - (2 \times 4\cdot2 \times 3\cdot7 \times \cos 110°)$

$c^2 = 17\cdot64 + 13\cdot69 - (^-10\cdot63)$

$c^2 = 41\cdot96$

$c = 6\cdot48 \text{cm}$

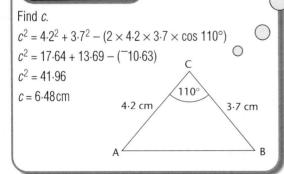

EXAMPLE 4

Find R.

$\cos R = \dfrac{6^2 + 8^2 - 5^2}{2 \times 6 \times 8}$

$\cos R = 0\cdot781\,25$

$R = \cos^{-1}(0\cdot781\,25)$

$R = 38\cdot6°$

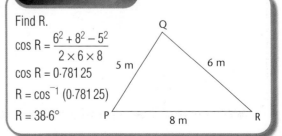

Chapter 5   *Trigonometry in non-right-angled triangles*

EXERCISE 5.2A

1

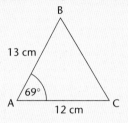

Find *a*.

2

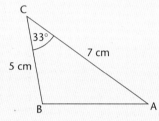

Find *c*.

3

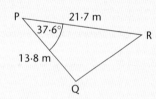

Find *p*.

4

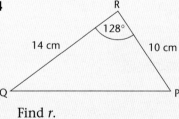

Find *r*.

5

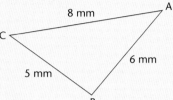

Find C.

6

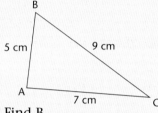

Find B.

7

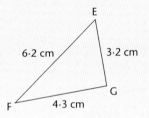

Find G.

8

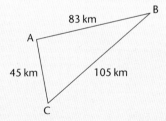

Three towns, A, B and C, are shown in the diagram. Find the angle ACB.

9 A cross-country runner runs 4 km due North and then 6·7 km in a South-East direction. How far is she from her starting point?

Exercise 5.2A cont'd

10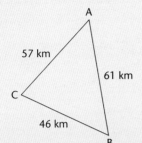

Three towns, A, B and C, are positioned as shown in the diagram. Find the three angles inside the triangle formed.

11 Three points, A, B and C, form an equilateral triangle on horizontal ground with sides 10 m.

Vertical posts PA, QB and RC are placed in the ground.

PA = 4 m, QB = 8 m and RC = 6 m.

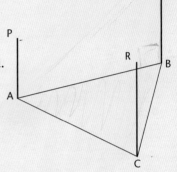

a) Calculate the lengths of:

(i) PQ

(ii) PR

(iii) QR.

b) Hence find the size of angle RPQ.

12 Use the data and result of question 11 to find the size of angle PRQ.

EXERCISE 5.2B

1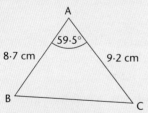

Find length BC.

2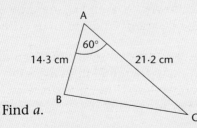

Find *a*.

3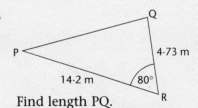

Find length PQ.

4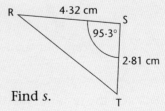

Find *s*.

Chapter 5 *Trigonometry in non-right-angled triangles*

Exercise 5.2B cont'd

5

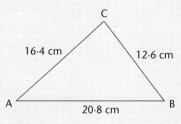

Find angle ABC.

6

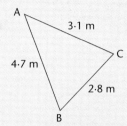

Find angle B.

7

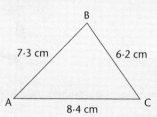

Find the smallest angle in the triangle.

8 From a boat, C, A is 9 km on a bearing of 058° and B is 12 km on a bearing of 110°. Calculate the distance AB.

9

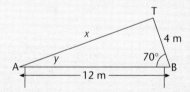

A roof structure is shown, spanning a house 12 m wide. Calculate the length of the roof, marked x, and the angle of the slope, $y°$.

10 A parallelogram has sides of length 5·1 cm and 2·5 cm. Adjacent sides are separated by an angle of 70°. Find the length of each of the diagonals of the parallelogram.

11

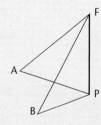

A vertical flagpole, FP, 25 m high, stands on horizontal ground. It is supported by two ropes, FA and FB. FA = 35 m and FB = 40 m. The angle between the ropes is 75°.

Work out the length AB.

12 Use the data and the result of question 11 to find the angle APB.

ACTIVITY 3

1   Just a two-step problem this time.
    Find the area of this triangle, ABC.

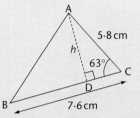

2   A much easier generalisation this time than in Activities 1 and 2
    a) Use triangle ACD to find *h* in terms of *b* and C.
    b) Use this expression for *h* to show that the area of triangle ABC is
    $\frac{1}{2}ab\sin C$.

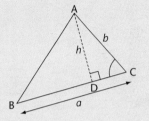

General formula for the area of any triangle

Once again, there are three versions of the formula and again
the letters have a 'circular' structure. Each formula requires
two adjacent sides and the included angle.

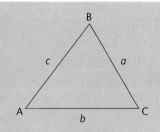

$$\text{Area of triangle ABC} = \frac{1}{2}ab\sin C$$
$$= \frac{1}{2}bc\sin A$$
$$= \frac{1}{2}ca\sin B$$

EXAMPLE 5

Find the area of the
triangle shown.

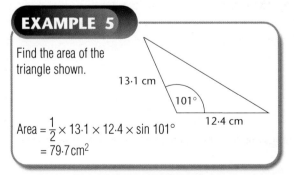

$\text{Area} = \frac{1}{2} \times 13 \cdot 1 \times 12 \cdot 4 \times \sin 101°$
    $= 79 \cdot 7\,\text{cm}^2$

Chapter 5   *Trigonometry in non-right-angled triangles*

EXERCISE 5.3A

Find the area of each of the following triangles.

1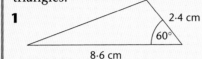

2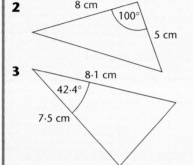

3

4 In triangle ABC, $a = 10\,cm$, $c = 6\,cm$ and B = 150°. Find the area of the triangle.

5 Calculate the area of parallelogram ABCD in which AB = 6 cm, BC = 9 cm and angle ABC = 41·4°.

EXERCISE 5.3B

Find the area of each of the following triangles.

1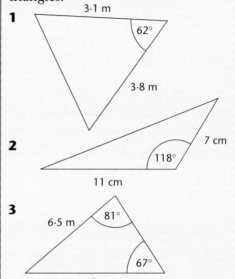

2

3

4 In triangle ABC, $a = 4\,cm$, $c = 7\,cm$ and its area is $13·4\,cm^2$. Find the size of angle B.

5 The area of triangle PQR is $273\,cm^2$. Given that PQ = 12·8 cm and angle PQR = 107°, find QR.

Key ideas

● To calculate lengths and angles in triangles without a right angle, use:
  – the sine rule when what you know is
  a) two angles and one side **b)** two sides and the non-included angle
  – the cosine rule when what you know is
  a) all the sides **b)** two sides and the included angle.

● To calculate the area of a triangle when you don't know the height, use area = $\frac{1}{2}ab \sin C$ or its equivalent.

Trends and time series

6

This chapter gives you more practice in interpreting time series and graphs modelling real-life situations.

EXAMPLE 1

The table shows the value, in thousands of pounds, of an ice-cream company's quarterly sales for 1995 to 1998. These sales have been plotted on the graph, together with the 4-quarter moving averages.

	1st quarter	2nd quarter	3rd quarter	4th quarter
1995	145	256	328	258
1996	189	244	365	262
1997	190	266	359	250
1998	201	259	401	265

> **Exam tip**
>
> Reminder – moving averages are plotted at the middle of the time period to which they refer.

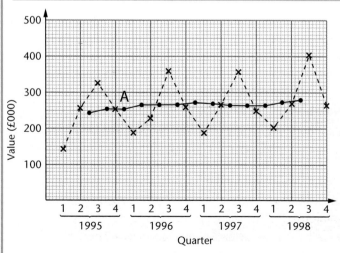

a) One of the moving average points has been marked A. Show how this point has been calculated.

b) Describe the trend in the moving averages and use this to forecast an estimate of the sales for the first quarter of 1999.

a) The point marked A is the moving average for quarters 3 and 4 of 1995 and 1 and 2 of 1996.

Its value $= \dfrac{328 + 258 + 189 + 244}{4}$

$= 254{\cdot}75$ thousand pounds.

Example 1 cont'd

b) The trend is a slight increase. The last moving average plotted is 281·5. Using an estimate of 285 for the next moving average gives a total of 1140 for these 4 quarters. The total sales for the last 3 quarters of 1998 are 925.

Estimate for first quarter of 1999 = 1140 – 925 = 215 thousand pounds.

ACTIVITY

Look on the internet to find other graphs showing time series and trends or seasonal variation.

Good sites to look at are those of government statistics showing social trends, or those with weather records.

Informal methods like that used in Example 1, part b) are often used to estimate future performance, but of course this may be affected by unknown factors. Do not rely too much on forecasts!

EXAMPLE 2

This graph shows the velocity of a car.

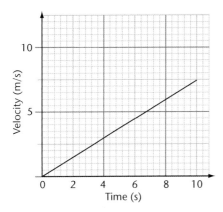

a) Calculate the gradient of this graph and say what it represents.

b) Sketch a graph to show the shape of the graph of the distance travelled by the car plotted against the time.

a) Gradient $= \dfrac{7.5\,\text{m/s}}{10\,\text{s}} = 0.75\,\text{m/s}^2$

This represents the acceleration of the car.

b) As the velocity increases, the car goes further in any given period of time.

The shape of the graph is this:

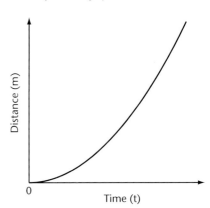

Exam tip

Always include the units when working out the gradients of graphs showing practical situations. The units help you to see what the gradient represents.

EXERCISE 6.1A

1 The table shows the daily sales figures in a clothing shop over a 4-week period.

	M	T	W	T	F	S	S
Week 1	1256	1785	875	2564	1932	3954	2703
Week 2	1307	2013	1076	2197	2102	4023	2893
Week 3	1293	1847	1132	1988	2234	4456	3215
Week 4	1402	1951	1207	2394	1987	5201	2694

a) Plot these figures on a graph.

b) Calculate the 7-day moving averages.

c) Plot the moving averages and add a trendline if appropriate.

d) Comment on any trend and daily variation.

2 a) Calculate the last moving average for this graph, which has not been plotted.

b) Comment on the seasonal variation and trends in the spending shown on this graph.

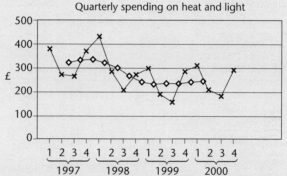

3 The table shows the daily audiences for the 4-week run of a play.

	Mon	Tues	Wed	Thur	Fri	Sat
Week 1	256	314	328	461	600	590
Week 2	307	341	392	421	600	562
Week 3	159	302	356	521	565	550
Week 4	173	261	403	488	524	481

a) Comment on the daily variation.

b) Calculate the 6-day moving averages.

c) Comment on the variation shown in the moving averages.

Exercise 6.1A cont'd

4

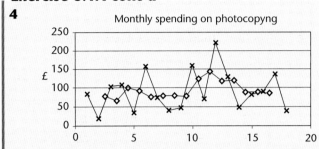

Monthly spending on photocopyng

Show how the highest point on the moving averages has been calculated. Explain what factors on the graph of monthly spending have caused this to be the highest point.

5

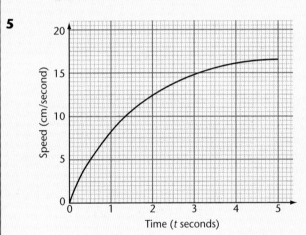

This graph shows the speed of a ball bearing falling through a cylinder filled with thick oil.

Describe what this graph shows about the acceleration of the ball bearing.

6

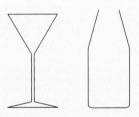

These two containers are filled at a constant rate. Sketch a graph for each one showing how the depth in the container varies with time.

7

The graph models water pouring into a bath. Describe what is happening at points A, B, C, D and E. Find the rate of flow at A.

EXERCISE 6.1B

1 The table shows the daily attendance figures at a cinema over a 4-week period.

	M	T	W	T	F	S	S
Week 1	137	201	414	216	521	823	411
Week 2	203	156	304	242	487	615	273
Week 3	166	231	322	284	534	725	389
Week 4	189	177	385	302	498	782	450

a) Plot these figures on a graph.

b) Calculate the 7-day moving averages.

c) Plot the moving averages and add a trendline if appropriate.

d) Comment on any trend and daily variation.

2 **a)** Show how the third moving average plotted on this graph has been calculated.

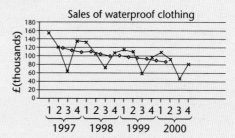

b) Comment on the seasonal variation and trends in the sales of waterproof clothing shown on this graph.

3 An insurance company gave these bonuses on the sums assured.

	1991	1992	1993	1994	1995	1996	1997	1998	1999	2000
Bonus (£%)	4·40	3·84	3·36	3·24	3·24	3·24	3·24	2·82	2·40	1·92

a) Comment on the trends.

b) Giving your reasons, forecast the bonus for 2001.

4 The table below shows the average temperature (in °C) for Westfield-on-Sea over a 5-day period. Calculate the 6-point moving averages and plot the graph. Comment on the weather pattern seen.

	12–4 a.m.	4–8 a.m.	8–12 p.m.	12–4 p.m.	4–8 p.m.	8–12 a.m.
Monday	⁻2	2	6	8	4	0
Tuesday	⁻1	2	8	7	5	3
Wednesday	2	3	7	9	5	2
Thursday	4	6	9	11	6	5
Friday	5	6	8	9	5	4

Chapter 6 *Trends and time series*

Exercise 6.1B cont'd

5

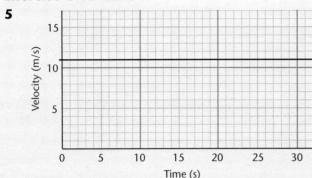

This graph shows a car moving with constant velocity.

Draw the graph of the distance it goes against time.

6

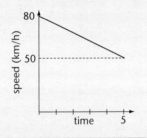

These two containers are emptied at a constant rate. Sketch a graph for each one showing how the depth in the container varies with time.

7

The graph shows a lorry travelling uphill on the motorway M172.

Calculate the deceleration of the lorry.

Key ideas

- A time series shows the variation of sets of figures over periods of time. These periods can be quarterly, daily, monthly, etc. These are usually displayed on a graph.
- To calculate a moving average, e.g. for quarterly figures, first calculate the mean for the first four quarters. Then omit the first quarter and include the fifth quarter and find the new mean. Then omit the second quarter and include the sixth and so on.
- The moving averages are plotted at the middle of the interval.

7 Congruency – proving and using

You should already know

- how to draw the perpendicular bisector of a line
- how to draw an angle bisector.

Congruent triangles

ACTIVITY 1

Work in pairs.

1 Each draw a triangle with sides 3 cm, 4 cm, 5 cm.

Are your triangles congruent?

2 Each draw a triangle, ABC, with AB = 6 cm, A = 60°, BC = 5·2 cm.

Are your triangles congruent?

If they are, can you draw one which isn't?

3 Try to find other examples of information given which can lead to:

a) one triangle only

b) two possible triangles.

For any two triangles to be congruent they must be the same shape and the same size. This means that they will fit exactly onto each other when one of them is rotated, reflected or translated.

Thus, it can be seen that two triangles will be congruent if:

- the three angles in one triangle equal the corresponding three angles of the other triangle
- the three sides in one triangle equal the corresponding three sides of the other triangle.

Two triangles are congruent if any of these conditions are satisfied:

- two sides and the included angle of one triangle are equal to the two sides and the included angle of the other triangle (side, angle, side or SAS)

- the three sides of one triangle are equal to the corresponding three sides of the other triangle (side, side, side or SSS)

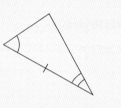 

- two angles and the side 'linking' them in one triangle are equal to the corresponding two angles and side in the other triangle (angle, side, angle or ASA)

- each triangle is right-angled and the hypotenuse and one side of one triangle is equal to the hypotenuse and a side of the other triangle (right angle, hypotenuse, side or RHS).

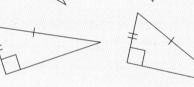

Note that the test invoving two sides and an angle is valid only if the angle is the included angle. If the given angle is not the included angle then there are two possible solutions – the ambiguous case which you may have met when studying the sine rule, or in Activity 1 in this chapter.

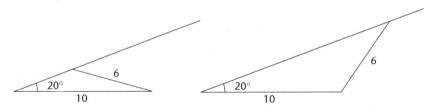

Chapter 7   *Congruency - proving and using*

EXAMPLE 1

Prove that the diagonal of a parallelogram splits the parallelogram into two congruent triangles. Hence show that the opposite sides of a parallelogram are equal.

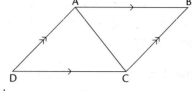

Angle BAC = angle ACD (alternate angles, AB and DC parallel).

Angle DAC = angle ACB (alternate angles, AD and BC parallel).

In triangles ACB and ACD, AC is common.

Hence these triangles are congruent (ASA).

Since the triangles are congruent, corresponding pairs of sides are equal.

So AD = CB

and AB = CD.

i.e. the opposite sides of a parallelogram are equal.

EXAMPLE 2

Use congruency to show that the reflection A'B' of a line AB in any line has a length equal to AB.

Let AA' and BB' cross the mirror line at C and D respectively

By definition of reflection, AC = A'C and AA' crosses the mirror line at right angles.

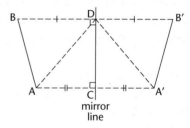

So considering triangles ACD and A'CD, since CD is common, these triangles are congruent SAS.

This implies AD = A'D and ∠ADC = ∠A'DC.

Considering triangles BDA and B'DA',
by definition of reflection BD = B'D and ∠BDC = ∠B'DC = 90°,
so ∠BDA = ∠B'DA' = 90° − ∠CDA.

So triangles BDA and B'DA' are congruent and, in particular, AB = A'B' as required.

> This is an example of a logical proof which builds up from basic definitions. You might like to try adapting the proof for when the mirror line crosses AB or goes through A or B.

Chapter 7   *Congruency - proving and using*

EXERCISE 7.1A

1 Which of these triangles are congruent? Give a reason or explanation for your answer.

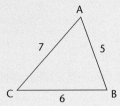
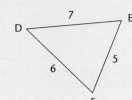
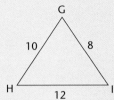
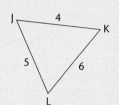

2 Which of these pairs of triangles are congruent? Explain your answer in each case.

a)

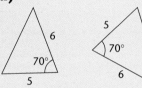

b)

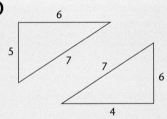

c)

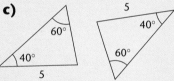

d)

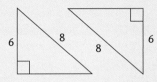

e)

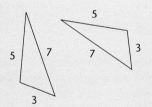

f)

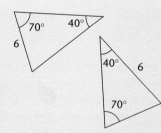

A sketch will prove useful for each question in the rest of this exercise.

3 Prove that the angles opposite the equal sides in an isosceles triangle are equal.

4 Prove that the diagonals of a rectangle are equal in length.

Exercise 7.1A cont'd

5 Sketch an isosceles triangle ABC with the equal angles at B and C. Draw a straight line from the mid point of side BC to the vertex A. Prove that this line bisects angle BAC and is perpendicular to side BC.

6 Use the property of congruent triangles to prove that the diagonals of a rhombus bisect each other at right angles.

7 Use the facts about a rhombus which you used in question 6 to explain why the construction to find the perpendicular bisector of a line works.

EXERCISE 7.1B

1 Which of the triangles **(i) (ii) (iii) (iv) (v) (vi)** are congruent to any of triangles A, B or C? Explain your answer in each case.

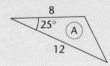
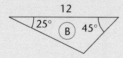
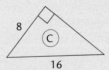

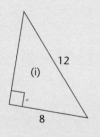
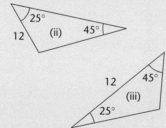
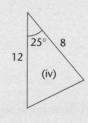

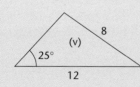
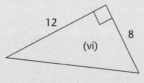

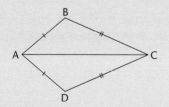

2 Prove that diagonal AC bisects the angles at A and C.

Chapter 7 *Congruency - proving and using*

Exercise 7.1B cont'd

A sketch will be useful for each question in the rest of this exercise.

3 Prove that the diagonals of a kite bisect the shorter diagonal at right angles and bisect the unequal pair of angles.

4 Show how what you proved in question 2 is used in the construction to bisect an angle.

5 Sketch an equilateral triangle. Join the midpoints of each side to make a second triangle. Prove that this triangle is equilateral.

6 Prove that the centre, O, of a rotation which maps AB onto A'B is where the perpendicular bisectors of AA' and BB' meet. Prove also that AB is equal in length to A'B'.

ACTIVITY 2

1 Use ruler and compasses to draw the perpendicular bisector of a line AB 10 cm long.

2 Mark any point on the bisector P. Join P to A and B.

3 Use congruent triangles to show why the construction works.

4 Use ruler and compasses to draw the bisector of angle ABC, which is 75°.

5 Use congruent triangles to show why the construction works.

Key ideas

Two triangles are congruent if any of these four sets of conditions is satisfied.

- The corresponding sides of each triangle are equal (SSS).
- Two sides and the included angle in each triangle are equal (SAS).
- Two angles and the corresponding side in each triangle are equal (ASA).
- Both triangles are right angled, the sides opposite the right angles are equal (i.e. the hypotenuse in each triangle) and another pair of sides are equal (RHS).

8 Calculating roots of equations

Reminder

The roots of an equation are the values for which it is true.

This quadratic equation factorises

$x^2 - 4x + 3 = 0$

This one does not

$x^2 - 4x + 1 = 0$

You have already learnt how to use graphs to solve equations like this, by drawing the graph of $y = x^2 - 4x + 1$ and reading off the values of x where $y = 0$.

In this chapter we look at two methods of calculating the roots of the equation.

ACTIVITY 1

Expand and simplify:

$(x + 1)^2 = (x + 1)(x + 1) =$

$(x - 2)^2 = (x - 2)(x - 2) =$

$(x + 3)^2 =$

$(x - 5)^2 =$

$(2x + 1)^2 =$

$(3x - 2)^2 =$

$(5x + 4)^2 =$

Spot the patterns in your answers and just write down the answer to these.

$(x - 4)^2$

$(4x + 1)^2$

$(2x - 3)^2$

ACTIVITY 2

Copy and fill in the missing spaces.

$x^2 + 4x + \quad = (x + 2)^2$

$x^2 + \quad + 9 = (x + 3)^2$

$x^2 + 10x + 25 = (x + \quad)^2$

$x^2 + 2x + \quad = (x + \quad)^2$

$x^2 + 12x + \quad = (x + \quad)^2$

$x^2 - 6x + \quad = (x - \quad)^2$

$x^2 - 16x + \quad = (x \quad)^2$

$x^2 - 7x + \quad = (x \quad)^2$

$x^2 + 5x + \quad = (x + \quad)^2$

$4x^2 + 12x + \quad = (2x + \quad)^2$

$9x^2 + 6x + \quad = (3x + \quad)^2$

$6x^2 + 24x + \quad = 6(x + \quad)^2$

Completing the square

$x^2 - 4x$ is part of the expansion $(x - 2)^2 = x^2 - 4x + 4$.

In this method for solving a quadratic equation we 'complete the square' so that the left-hand side of the equation is a square $(mx + p)^2$.

EXAMPLE 1

Solve $x^2 - 4x + 1 = 0$

$x^2 - 4x + 1 = 0$

$\quad\quad +3 \quad +3$

$x^2 - 4x + 4 = 3$ | Add a number so the left-hand side is a complete square.

$(x - 2)^2 = 3$ | Take the square root of both sides.

$x - 2 = \sqrt{3} \text{ or } {}^-\sqrt{3}$

$\quad\quad x = 2 + \sqrt{3} \text{ or } 2 - \sqrt{3}$ | This is usually written $x - 2 = \pm\sqrt{3}$

$x = 3{\cdot}75 \text{ or } 0{\cdot}27 \text{ correct to 2 d.p.}$

EXAMPLE 2

Solve $x^2 + 3x - 5 = 0$.

$x^2 + 3x + \dfrac{9}{4} = 5 + \dfrac{9}{4}$

$\left(x + \dfrac{3}{2}\right)^2 = \dfrac{29}{4}$

$x + \dfrac{3}{2} = \pm\sqrt{\dfrac{29}{4}}$

$x = \dfrac{{}^-3}{2} \pm \sqrt{\dfrac{29}{4}}$

$x = 1{\cdot}19 \text{ or } {}^-4{\cdot}19 \text{ to 2 d.p.}$

An alternative method to avoid fractions is to multiply the equation through by 4 at the start.

Look again at the equation in example 2.

$\quad\quad x^2 + 3x - 5 = 0$

$\quad 4x^2 + 12x - 20 = 0$ | Multiply by 4

$\quad\quad 4x^2 + 12x + 9 = 29$ | Add 29 so LHS is a complete square

$\quad\quad\quad (2x + 3)^2 = 29$

$\quad\quad\quad\quad 2x + 3 = \pm\sqrt{29}$ | Take the square root

$\quad\quad\quad\quad\quad 2x = {}^-3 \pm \sqrt{29}$ | Rearrange to make x the subject

$\quad\quad\quad\quad\quad\quad x = \dfrac{{}^-3 \pm \sqrt{29}}{2}$

$\quad\quad\quad\quad\quad\quad\quad = 1{\cdot}19 \text{ or } {}^-4{\cdot}19 \text{ to 2 d.p.}$

> **Hint**
>
> Use this method to work out how much to add:
> Since $(mx + k)^2 = m^2x^2 + 2kmx + k^2$ use the x^2 coefficient to see $m = 2$ then use the x coefficient to see $2km = 12$, so $k = 3$.
> We need $k^2 = 9$ on LHS.

If the coefficient of x^2 is not a square already, multiply the equation through at the start, to make it so.

EXAMPLE 3

Solve $2x^2 + 10x + 5 = 0$

Multiply by 2:

$$4x^2 + 20x + 10 = 0$$

$$4x^2 + 20x + 25 = 15 \qquad \text{Adding 15 to get 25 on LHS.}$$

$$(2x + 5)^2 = 15$$

$$(2x + 5) = \pm\sqrt{15}$$

$$2x = {}^-5 \pm \sqrt{15}$$

$$x = \frac{{}^-5 \pm \sqrt{15}}{2}$$

$$= {}^-0 \cdot 56 \text{ or } {}^-4 \cdot 44 \text{ to 2 d.p.}$$

EXERCISE 8.1A

Solve these equations by completing the square.

1 $x^2 - 6x + 4 = 0$

2 $x^2 - 8x - 2 = 0$

3 $x^2 + 10x + 5 = 0$

4 $x^2 + 3x - 1 = 0$

5 $x^2 - 7x + 2 = 0$

6 $4x^2 - 6x + 1 = 0$

7 $3x^2 - 4x - 5 = 0$

8 $2x^2 - 5x - 2 = 0$

9 $5x^2 - x - 1 = 0$

10 $16x^2 + 12x + 1 = 0$

EXERCISE 8.1B

Solve these equations by completing the square.

1 $x^2 - 2x - 2 = 0$

2 $x^2 + 6x - 4 = 0$

3 $x^2 + 8x + 3 = 0$

4 $x^2 - 10x + 6 = 0$

5 $x^2 + 3x - 5 = 0$

6 $x^2 - 5x + 2 = 0$

7 $4x^2 + 3x - 2 = 0$

8 $2x^2 + 12x + 3 = 0$

9 $3x^2 + 2x - 2 = 0$

10 $5x^2 - 6x - 4 = 0$

Using the quadratic formula

Using the method of completing the square it may be shown that the equation
$ax^2 + bx + c = 0$

has roots $x = \dfrac{{}^-b \pm \sqrt{b^2 - 4ac}}{2a}$

ACTIVITY 3

Try proving this formula for yourself.

Hint: multiply through to avoid fractions

If you are not specifically asked to use another method, just to solve the equation, you may quote and use this formula.

EXAMPLE 4

Solve the equation $x^2 - 4x + 2 = 0$.
Give your answer to 2 d.p.

$a = 1, b = {}^-4, c = 2$

$x = \dfrac{+4 \pm \sqrt{16 - 4 \times 1 \times 2}}{2}$

$= \dfrac{+4 \pm \sqrt{8}}{2}$

$= \dfrac{4 \pm 2 \cdot 828}{2} = \dfrac{6 \cdot 828}{2} \text{ or } \dfrac{1 \cdot 172}{2}$

$= 3 \cdot 414 \text{ or } 0 \cdot 586$

$= 3 \cdot 41 \text{ or } 0 \cdot 59 \text{ to 2 d.p.}$

> **Exam tip**
>
> Make sure you always include the correct sign when substituting for a, b and c.

> **Exam tip**
>
> It is sensible to do all the working out on a calculator, but it is vital to write down the expression you are going to work out.

EXAMPLE 5

Solve the equation $3x^2 + 4x - 2 = 0$.
Give the answers to 2 d.p.

$a = 3, b = 4, c = {}^-2$

$x = \dfrac{{}^-4 \pm \sqrt{16 - 4 \times 3 \times ({}^-2)}}{6} = \dfrac{{}^-4 \pm \sqrt{16 + 24}}{6} = \dfrac{{}^-4 \pm \sqrt{40}}{6}$

$= \dfrac{{}^-4 + 6 \cdot 324}{6} \text{ or } \dfrac{{}^-4 - 6 \cdot 324}{6}$

$= \dfrac{2 \cdot 324}{6} \text{ or } \dfrac{{}^-10 \cdot 324}{6} = 0 \cdot 3873 \text{ or } {}^-1 \cdot 7206$

$x = 0 \cdot 39 \text{ or } {}^-1 \cdot 72 \text{ to 2 d.p.}$

> **Exam tip**
>
> The main errors that occur in using the formula are:
> (1) errors with the signs, especially with ${}^-4ac$
> (2) failure to divide everything by $2a$ and not just the first term.

EXERCISE 8.2A

Use the formula to solve these equations. Give the answers correct to 2 d.p.

1 $x^2 + 8x + 6 = 0$

2 $2x^2 - 2x - 3 = 0$

3 $3x^2 + 5x - 1 = 0$

4 $5x^2 - 12x + 5 = 0$

5 $5x^2 + 9x - 6 = 0$

6 $x^2 - 5x - 1 = 0$

7 $3x^2 + 9x + 5 = 0$

8–10 Solve the equations in Exercise 8.1A questions 8–10 but this time using the formula.

11 A 'garden' is 8 m longer than it is wide and it has an area of 25 m². Write down an equation and solve it to find the dimensions correct to the nearest cm.

12 Dad is three times older than Tom. In five years' time Tom works out that the product of their ages will be roughly 1000. Write down an equation and solve it to find out how old they are now.

EXERCISE 8.2B

Use the formula to solve these equations. Give the answers correct to 2 d.p.

1 $x^2 + 7x + 4 = 0$

2 $2x^2 - 3x - 4 = 0$

3 $3x^2 + 2x - 2 = 0$

4 $5x^2 - 13x + 7 = 0$

5 $5x^2 + 9x + 3 = 0$

6 $7x^2 - 5x - 1 = 0$

7 $3x^2 + 2x - 7 = 0$

8–10 Solve the equations in Exercise 8.1B questions 8–10, but this time using the formula.

11 A pen is constructed along an existing wall using 20 m of fencing.

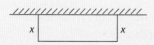

a) If the width is x m, write down an expression for the area enclosed.

b) Write down an equation and solve it to find the dimensions to give an area of 40 m².

c) By completing the square, find the maximum area possible with this amount of fencing.

Exercise 8.2B cont'd

12 A rectangular lawn measuring 22 m by 15 m is surrounding
  by a path x m wide. Form a simplified expression for the
  total area of lawn and path. Write down an equation and
  solve it to find the width of path correct to the nearest cm
  if the area is 400 m².

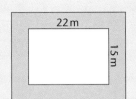

Key ideas

To calculate the roots of a quadratic equation which will not factorise:

● complete the square

  a) multiply if necessary so the coefficient of x^2 is a perfect square

  b) add a number to both sides so the left-hand side is a complete square

  c) take the square root of both sides

  d) solve the resulting linear equations

  or

● use the formula

  when $ax^2 + bx + c = 0$

  $$x = \frac{^-b \pm \sqrt{b^2 - 4ac}}{2a} .$$

Revision exercise

1 In triangle PRT, find:

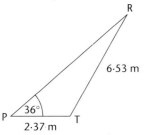

a) angle PRT

b) angle PTR

c) PR.

2 In triangle ABC, find:

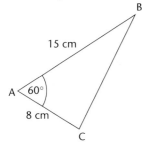

a) BC

b) angle ABC

c) the area of triangle ABC.

3 In triangle KLM, find:

a) angle LKM

b) angle KML.

4 In triangle ABC, find:

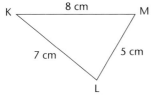

a) AC

b) angle BAC.

5 ABCD is a field with dimensions as shown in the diagram. Calculate the area of the field.

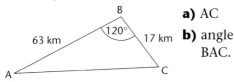

6 The figures in the table show Colin's electricity bills (in £).

	1st Qtr	2nd Qtr	3rd Qtr	4th Qtr
1995	120·34	78·61	56·98	110·55
1996	126·92	75·03	55·09	120·81
1997	132·67	81·32	61·14	123·50
1998	143·84	79·89	70·83	125·16

a) Calculate a four-point moving average and plot a graph.

b) What do you notice?

c) Can you predict the bills for the next four quarters?

7 The table gives the rainfall in millimetres for each month in Huangogo in Central Africa during a three-year period. The Africans refer to a rainy season and a dry season.

a) Identify when these are.

b) Calculate a suitable moving average and comment on the graph that you get.

	J	F	M	A	M	J
1995	25	40	67	104	15	2
1996	30	38	80	116	10	0
1997	28	46	91	115	18	6

	J	A	S	O	N	D
1995	0	0	0	4	29	21
1996	0	0	6	7	36	19
1997	0	0	4	11	40	23

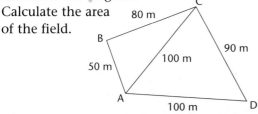

8 The graph shows the temperature of a commercial oven used for baking large quantities of pastry at one time.

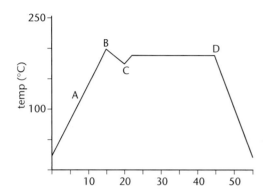

a) Describe what is happening at points A, B, C, and D.

b) Find the rate of warming up at A.

9 Join the mid points of the sides of a square to form a quadrilateral. Prove that this quadrilateral is a square.

10

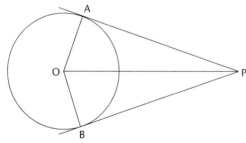

PA and PB are tangents to the circle with centre O. Prove that triangles AOP and BOP are congruent.

11 In the diagram for question 10, let D be the point where AB crosses OP. Prove that triangles PAD and PBD are congruent.

12 Solve these quadratic equations by completing the square. Give your answers correct to 2 d.p.

a) $x^2 - 6x + 2 = 0$

b) $4x^2 - 5x - 3 = 0$

c) $3x^2 + 4x - 2 = 0$

13 Write $y = x^2 - 5x + 4$ in the form $y = (x - a)^2 + b$. Hence state the coordinates of the minimum point on the graph of $y = x^2 - 5x + 4$.

14 Use the quadratic formula to solve these equations. Give your answers correct to 3 d.p.

a) $5x^2 - 8x + 1 = 0$

b) $x^2 - 7x - 2 = 0$

c) $6x^2 + 2x - 7 = 0$

9 Surface areas and complex shapes

You should already know

- how to find the circumference and area of circles
- how to find the arc length and area of a sector
- how to find the area of a triangle
- how to find the volume of a prism, pyramid, cone or sphere
- how to rearrange formulae
- how to use Pythagoras' theorem and trigonometry.

Surface areas

ACTIVITY

1   Make a 3D sketch of a rectangular-based pyramid.

What do you need to know to work out its volume?

2   Sketch the net of the same pyramid.

What do you need to know to work out its surface area? Can you work it out knowing the base measurements and the height?

Think of the label around a cylindrical can. It can be opened out flat to form a rectangle.

The length of the rectangle is the circumference of the can. Its width is the height of the can. The area of the rectangle is the curved surface area of the cylinder.

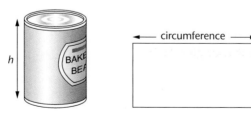

This gives: curved surface area of cylinder = $2\pi rh$.

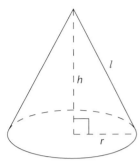

The curved surface area of a cone can be opened out to form the sector of a circle of radius l, where l is the slant height of the cone. The arc length of the sector is the circumference of the base of the cone.

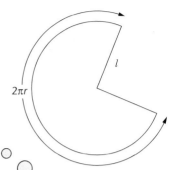

You might like to use the work on sectors that you did in Stage 9 to prove the formula for yourself.

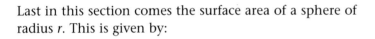

Curved surface area of cone = $\pi r l$.

Last in this section comes the surface area of a sphere of radius r. This is given by:

surface area of sphere = $4\pi r^2$.

> **Exam tip**
>
> Make sure you distinguish between the perpendicular height h and the slant height l of a cone – and don't read l as 1!

EXAMPLE 1

This cone has a solid base. Find:

a) the volume of the cone

b) the total surface area of the cone.

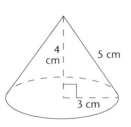

a) Volume $= \frac{1}{3}\pi \times 3^2 \times 4 = 12\pi = 37{\cdot}7\,\text{cm}^3$ to 3 s.f.

b) Curved surface area $= \pi \times 5 = 15\pi$

Area of base $= \pi \times 3^2 = 9\pi$

Total surface area $= 24\pi = 75{\cdot}4\,\text{cm}^2$ to 3 s.f.

EXERCISE 9.1A

1 Calculate the curved surface area of these cylinders:

a) radius = 4·7 cm, height = 8·2 cm

b) radius = 1·2 m, height = 2·5 m

c) radius = 3·5 cm, height = 4·6 cm.

Exercise 9.1A cont'd

2 Calculate the curved surface area of these cones.

a)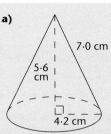

b)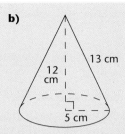

c)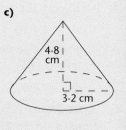

3 Calculate the surface area of a sphere of radius: **a)** 5 cm **b)** 6·2 cm **c)** 2 mm.

4 What is the total surface area of this wooden cylindrical brick?

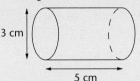

5 A sphere has a surface area of 47·6 cm². Calculate its radius.

6 A solid cone has a base radius of 4·5 cm and a height of 6·3 cm. Calculate its total surface area.

7 Calculate the slant height of a cone of base radius 5 cm and surface area 120 cm².

8 A solid cylinder of length 6·0 cm has a curved surface area of 1800 cm². Calculate its radius.

9 A cylinder has height 7·2 cm and base area 36 cm². Calculate its curved surface area.

10 The flat surface of a hemisphere has an area of 85 cm². Calculate its curved surface area.

EXERCISE 9.1B

1 Calculate the curved surface area of these cylinders:

a) radius = 2·7 cm, height = 3·4 cm   **b)** radius = 1·9 m, height = 1·6 m

c) radius = 7·2 cm, height = 15·7 cm.

2 Calculate the curved surface area of these cones.

a)

b)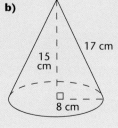

c)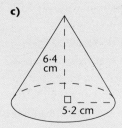

Chapter 9 *Surface areas and complex shapes*

Exercise 9.1B cont'd

3 Calculate the surface area of a sphere of radius:

    a) 3 cm   **b)** 4·7 cm   **c)** 5 mm.

4 Calculate the total surface area of a solid cone with base radius 7·1 cm and slant height 9·7 cm.

5 A sphere has a surface area of 157·6 cm^2. Calculate its radius.

6 A wastepaper bin is a cylinder with no lid. It is 50 cm high and the diameter of its base is 30 cm. The outside of the bin is painted white; the inside is painted black. Calculate the area that is painted black.

7 A cylinder 6·3 cm long has a curved surface area of 170 cm^2. Calculate the radius of the cylinder.

8 Calculate the base area of a cone with slant height 8·2 cm and curved surface area 126 cm^2.

9 Calculate the total surface area of a solid hemisphere of radius 5·2 cm.

10 A cone has slant height 7 cm and curved surface area 84 cm^2. Calculate the total surface area of the cone.

More complex problems

These may involve combining shapes you have met earlier. Or they may use topics you have met in other chapters, such as Pythagoras' theorem and trigonometry. Be prepared for anything, and enjoy the problem-solving!

One shape you may not have met before is the **frustum** of a cone. This is the shape remaining when a solid cone has a smaller cone removed from it as shown in this diagram.

Remove the top cone.

This shape is a frustum.

Exam tip

When dealing with problems where you have to work out how to solve them, follow these steps:

Read the question carefully and plan:
- What do I know?
- What do I have to find?
- What methods can I apply?

Look back when you have finished and ask 'Have I answered the question?' There may be one last step you have forgotten to do.

EXAMPLE 2

Calculate the area of this segment.

Area of segment = area of sector − area of triangle.

Area of sector $= \dfrac{100}{360} \times \pi \times 5^2$

$= 21\,816...\,\text{cm}^2$

Area of triangle $= \dfrac{1}{2}\,ab\,\sin C$

$= \dfrac{1}{2} \times 5^2 \times \sin 100°$

$= 12·31...\,\text{cm}^2$

Area of segment $= 21·816... − 12·31...$

$= 9·51\,\text{cm}^2$ to 3 s.f.

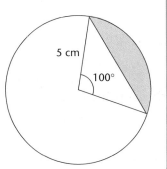

Exam tip

Write down more figures than you need in the working, and round the final answer. Using the calculator memory means you do not have to re-key figures.

EXAMPLE 3

Calculate the curved surface area of this cone.

First the slant height l must be found.

$l^2 = 5^2 + 6^2 = 51$ $l = \sqrt{51}$

Curved surface area $= \pi r l$

$= \pi \times 5 \times \sqrt{51}$

$= 112\,\text{cm}^2$ to 3 s.f.

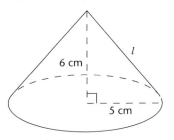

EXERCISE 9.2A

1   A solid cylinder has a base radius of 3 cm. Its volume is 95 cm³. Calculate its curved surface area.

2

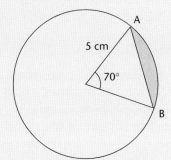

Calculate:

a)   the length of the chord AB

b)   the perimeter of the shaded segment.

Exercise 9.2A cont'd

3

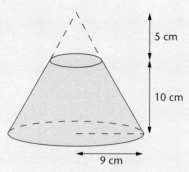

5 cm

10 cm

9 cm

A cone of height 15 cm and base 9 cm has a cone of height 5 cm removed from its top as shown.

a) What is the radius of the base of the top cone?

b) Calculate the volume of the remaining frustum of the cone.

4

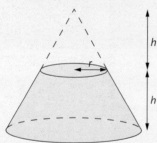

h

r

h

Show that the volume of the frustum is $\frac{7}{3}\pi r^2 h$.

5 All the edges of this square-based pyramid are 5 cm. Calculate:

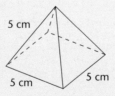

5 cm

5 cm

5 cm

a) its perpendicular height

b) its volume.

6 a) Show that the area of this shaded segment may be expressed as

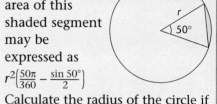

r

50°

$r^2\left(\frac{50\pi}{360} - \frac{\sin 50°}{2}\right)$

b) Calculate the radius of the circle if the area of the segment is 2 cm².

7 A cone has height 6 cm and volume 70 cm³.

a) Calculate its base radius.

b) Calculate its curved surface area.

8 A paintball sphere has capacity of 1 litre. Calculate the surface area of the sphere.

9 Calculate the area of the major shaded segment in this diagram.

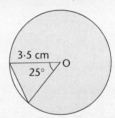

3·5 cm

O

25°

> **Exam tip**
>
> Capacity is the amount (volume) that a container can hold.

10

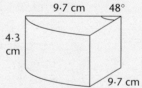

9.7 cm   48°

4·3 cm

9.7 cm

A piece of cheese is a prism whose cross-section is the sector of a circle with measurements as shown. Calculate the volume of the piece of cheese.

EXERCISE 9.2B

1

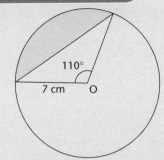

Calculate the area of this shaded segment.

2

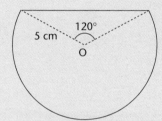

Calculate the perimeter of this segment of a circle of radius 5 cm.

3 The bowl of this glass is part of a sphere. The radius of the sphere is 5 cm. The radius of the top of the glass is 3·7 cm. Calculate the depth, d, of the glass.

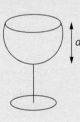

4

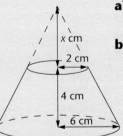

a) Explain why $x + 4 = 3x$.

b) Calculate the volume of the frustum.

5 A cone has a base of radius 5 cm and perpendicular height of 12 cm. Calculate its curved surface area.

6 A cone's perpendicular height is equal to its base radius. The volume of the cone is 24 cm^3. Calculate its perpendicular height.

7

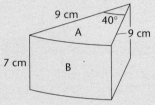

The faces labelled A and B of this slice of cake are covered in chocolate. The complete cake is a cylinder of radius 9 cm and depth 7 cm. What area of the slice is covered in chocolate?

8 A spherical ball has a curved surface area of 120 cm^2. Calculate its volume.

9

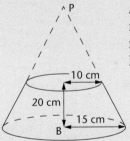

A lampshade is made from the frustum of a hollow cone.

a) Show that the slant height, AP, of the complete cone is $15\sqrt{17}$ cm.

b) Calculate the curved surface area of the lampshade.

10 All the sloping edges of this square-based pyramid are 8 cm long. Calculate the volume of the pyramid.

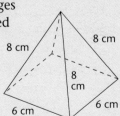

Key ideas

- Arc length $= \frac{\theta}{360} \times 2\pi r$.
  Sector area $= \frac{\theta}{360} \times \pi r^2$.

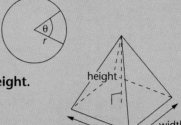

- Volume of pyramid $= \frac{1}{3}$ length × width × height.

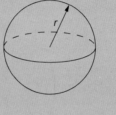

- Volume of sphere $= \frac{4}{3}\pi r^3$.
  Surface area of sphere $= 4\pi r^2$.

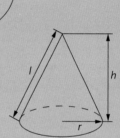

- Volume of cone $= \frac{1}{3}\pi r^3$.
  Curved surface area of cone $= \pi r l$.

- The frustum of a cone is the shape remaining when a solid cone has a smaller cone removed from it as shown in this diagram.

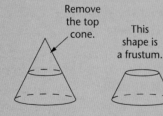

Remove the top cone.

This shape is a frustum.

- Area of segment = area of sector − area of triangle.

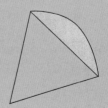

Chapter 9 *Surface areas and complex shapes*

10 Working with algebraic fractions

You should already know

- how to add, subtract, multiply, divide and simplify numeric fractions
- how to simplify and factorise algebraic expressions
- how to rearrange formulae.

ACTIVITY

1 Remind yourself how to add and subtract fractions.

a) $\frac{2}{3} + \frac{5}{8}$ b) $\frac{3}{4} + \frac{7}{20}$

c) $\frac{9}{8} - \frac{2}{15}$ d) $\frac{x}{3} + \frac{5x}{8}$

e) $\frac{5x}{9} - \frac{4x}{15}$

2 Write these fractions in their lowest terms.

a) $\frac{5}{20}$ b) $\frac{16}{24}$

c) $\frac{3x}{9}$ d) $\frac{6}{27x}$

e) $\frac{9a}{6a^2}$

Algebraic fractions

When an expression involving fractions needs to be simplified, it should be put over a common denominator.

EXAMPLE 1

Simplify $\frac{x+3}{2} - \frac{x-3}{3}$

The common denominator is 6, so $(x + 3)$ is multiplied by $6 \div 2 = 3$ and $(x - 3)$ is multiplied by $6 \div 3 = 2$.

$$\frac{x+3}{2} - \frac{x-3}{3} = \frac{3(x+3) - 2(x-3)}{6} = \frac{3x+9 - 2x + 6}{6}$$

$$= \frac{x+15}{6}$$

If the denominators involve x the procedure is still the same.

EXAMPLE 2

Simplify $\frac{3}{x+1} - \frac{2}{x}$

The common denominator is $x(x + 1)$, so 3 is multiplied by $x(x +1) \div (x + 1) = x$ and 2 is multiplied by $x(x + 1) \div x = (x + 1)$.

$$\frac{3}{x+1} - \frac{2}{x} = \frac{3x - 2(x+1)}{x(x+1)} = \frac{3x - 2x - 2}{x(x+1)} = \frac{x-2}{x(x+1)}$$

Exam tip

Errors often occur through cancelling individual terms. Only factors, which can be individual numbers, letters or brackets, can be cancelled.

Exam tip

Do not miss out the first step. Most errors occur because of expanding the brackets wrongly without writing them down.

205

EXAMPLE 3

Simplify $\dfrac{x}{x+3} - \dfrac{x-2}{x} + \dfrac{2}{5}$.

The common denominator is $5x(x+3)$. Multiply x by $5x$, $(x-2)$ by $5(x+3)$ and 2 by $x(x+3)$.

$$\frac{x}{x+3} - \frac{x-2}{x} + \frac{2}{5} = \frac{5x^2 - 5(x+3)(x-2) + 2x(x+3)}{5x(x+3)} = \frac{5x^2 - 5(x^2 + x - 6) + 2x^2 + 6x}{5x(x+3)}$$

$$= \frac{5x^2 - 5x^2 - 5x + 30 + 2x^2 + 6x}{5x(x+3)} = \frac{2x^2 + x + 30}{5x(x+3)}$$

EXERCISE 10.1A

Simplify:

1 $\dfrac{x}{2} + \dfrac{3x}{5}$

2 $\dfrac{x+1}{3} - \dfrac{2x-1}{2}$

3 $\dfrac{x-3}{5} + \dfrac{2x}{3} - \dfrac{3x-2}{10}$

4 $\dfrac{1}{x} + \dfrac{2}{x-1}$

5 $\dfrac{3}{2x} - \dfrac{1}{x+2}$

6 $\dfrac{2}{x+1} + \dfrac{3}{x-1}$

7 $\dfrac{2x}{3x+1} - \dfrac{5}{x+3}$

8 $\dfrac{x+1}{x-1} + \dfrac{3x-1}{x+2}$

9 $\dfrac{x}{x+1} - \dfrac{3}{5} + \dfrac{x-2}{x}$

10 $\dfrac{2x}{x-3} + \dfrac{x-1}{x+2} - \dfrac{4}{9}$

EXERCISE 10.1B

Simplify:

1 $\dfrac{2x}{3} - \dfrac{3x}{5}$

2 $\dfrac{x-1}{2} - \dfrac{x-3}{5}$

3 $\dfrac{2x-1}{6} + \dfrac{3x}{4} - \dfrac{x-2}{12}$

4 $\dfrac{3}{x} + \dfrac{2}{x+1}$

5 $\dfrac{5}{6x} - \dfrac{1}{2x+1}$

6 $\dfrac{3}{x+2} + \dfrac{5}{x-1}$

7 $\dfrac{2x}{x+1} - \dfrac{x-1}{x+3}$

8 $\dfrac{2x}{x-1} - \dfrac{3x+2}{x+2}$

9 $\dfrac{x}{x+1} + \dfrac{3}{5} - \dfrac{x+3}{x}$

10 $\dfrac{2}{x-1} - \dfrac{3}{x+2} - \dfrac{1}{x+3}$

Harder equations

You need to be able to solve equations involving brackets and fractions. These may be linear or quadratic.

Exam tip

In equations with algebriac fractions multiply through by the common denominator. This removes the fractions.

EXAMPLE 4

Solve $\dfrac{3}{x+1} - \dfrac{2}{x} = \dfrac{1}{x-2}$.

Multiply by $x(x+1)(x-2)$. Make sure you multiply every expression on both sides.

The equation becomes $3x(x-2) - 2(x+1)(x-2) = x(x+1)$

$$3x^2 - 6x - 2(x^2 - x - 2) = x^2 + x$$

$$3x^2 - 6x - 2x^2 + 2x + 4 - x^2 - x = 0 \qquad \text{Expanding out and collecting on one side.}$$

$$^-5x + 4 = 0 \qquad \text{Collect like terms.}$$

$$5x = 4 \qquad \text{Rearrange and change signs.}$$

$$x = \frac{4}{5}$$

EXAMPLE 5

Solve $(x-1)^2 = 7 - x$.   The first thing to do is multiply out the bracket.

The equation becomes:

$$x^2 - 2x + 1 = 7 - x$$

Remember
$(x-1)^2 = (x-1)(x-1)$.

$$x^2 - x - 6 = 0$$

Collect all terms on one side.

$$(x-3)(x+2) = 0$$

$$x = 3 \text{ or } ^-2.$$

EXAMPLE 6

Solve $\dfrac{5}{x} = \dfrac{x}{x+10}$.   Multiply through by $x(x+10)$.

The equation becomes:

$$5(x+10) = x^2$$

$$5x + 50 = x^2 \qquad \text{Expand bracket.}$$

$$^-x^2 + 5x + 50 = 0 \qquad \text{Collect all terms on one side.}$$

$$x^2 - 5x - 50 = 0 \qquad \text{Change all signs.}$$

$$(x-10)(x+5) = 0$$

$$x = 10 \text{ or } ^-5$$

EXERCISE 10.2A

Solve:

1. $\dfrac{2x}{3} - \dfrac{3x}{5} = \dfrac{1}{3}$

2. $\dfrac{x-1}{2} - \dfrac{x-3}{5} = 1$

3. $\dfrac{2x-1}{6} + \dfrac{3x}{4} = \dfrac{x-2}{12}$

4. $\dfrac{3}{x} - \dfrac{2}{x+1} = 0$

5. $\dfrac{5}{6x} - \dfrac{1}{x+1} = \dfrac{1}{3x}$

6. $x(x+2) = 2(x+2)$

7. $2x(x-2) = x^2 + 5$

8. $4x = \dfrac{3}{x} - 1$

9. $2x^2 - \dfrac{x}{3} = 5$

10. $\dfrac{1}{x-1} - \dfrac{3}{x+2} = \dfrac{1}{4}$

EXERCISE 10.2B

Solve:

1. $\dfrac{2x}{5} - \dfrac{x}{4} = \dfrac{3}{10}$

2. $\dfrac{2x-1}{2} - \dfrac{x-3}{3} = \dfrac{5}{2}$

3. $\dfrac{x-1}{3} + \dfrac{2x}{5} = \dfrac{3x+1}{5}$

4. $\dfrac{5}{x} - \dfrac{2}{x-3} = 0$

5. $\dfrac{4}{x-2} - \dfrac{1}{x+1} = \dfrac{3}{x}$

6. $(x+4)(x+2) + x + 4 = 0$

7. $(x-5)(x+3) = x - 5$

8. $4x + \dfrac{3}{x} = 7$

9. $2x + \dfrac{4}{x} = 9$

10. $\dfrac{2x}{3x+1} - \dfrac{5}{x+3} = 0$

Key ideas

- When adding or subtracting fractions, put them over a common denominator.
- When cancelling algebraic fractions, factorise if necessary. Only cancel factors.
- When equations involve fractions, multiply through by the common denominator to remove the fractions.

Chapter 10 *Working with algebraic fractions*

11 Vectors

You should already know

- how to add, subtract, multiply, divide and simplify numeric fractions
- how to simplify and factorise algebraic expressions
- how to rearrange formulae.

Column vectors

If a vector is drawn on a coordinate grid then it can be described by a column vector $\begin{pmatrix} x \\ y \end{pmatrix}$, where x is the length across to the right and y is the length upwards.

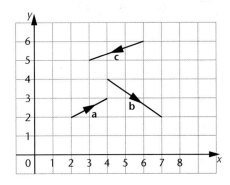

In the diagram $\mathbf{a} = \begin{pmatrix} 2 \\ 1 \end{pmatrix}$ $\mathbf{b} = \begin{pmatrix} 3 \\ -2 \end{pmatrix}$ $\mathbf{c} = \begin{pmatrix} -3 \\ -1 \end{pmatrix}$

EXAMPLE 1

Write down the column vectors for
$\overrightarrow{AB}$, $\overrightarrow{BC}$, $\overrightarrow{CD}$, $\overrightarrow{AD}$, $\overrightarrow{BD}$, $\overrightarrow{DC}$.

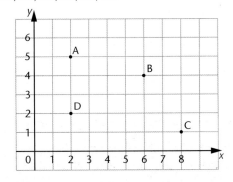

$\overrightarrow{AB} = \begin{pmatrix} 4 \\ -1 \end{pmatrix}$, $\overrightarrow{BC} = \begin{pmatrix} 2 \\ -3 \end{pmatrix}$, $\overrightarrow{CD} = \begin{pmatrix} -6 \\ 1 \end{pmatrix}$, $\overrightarrow{AD} = \begin{pmatrix} 0 \\ -3 \end{pmatrix}$,

$\overrightarrow{BD} = \begin{pmatrix} -4 \\ -2 \end{pmatrix}$, $\overrightarrow{DC} = \begin{pmatrix} 6 \\ -1 \end{pmatrix}$.

> ## Exam tip
>
> Column vectors must be columns. If you write them down as coordinates it will be marked as wrong.

> ## Exam tip
>
> In examinations, candidates often make an error of 1 when working out the values for the vector. Take care with the counting.

ACTIVITY 1

Draw a grid with both axes from ‾6 to 6.

Anywhere on the grid, draw these column vectors: $\begin{pmatrix}1\\3\end{pmatrix}$ $\begin{pmatrix}2\\6\end{pmatrix}$ $\begin{pmatrix}3\\9\end{pmatrix}$.

What do you notice?

Repeat with these vectors: $\begin{pmatrix}3\\2\end{pmatrix}$ $\begin{pmatrix}6\\4\end{pmatrix}$ $\begin{pmatrix}9\\6\end{pmatrix}$

General vectors

A vector has both length and direction but can be in any position. The vector going from A to B can be labelled $\overrightarrow{AB}$ or or it can be given a letter **a**, heavy type. When handwritten, put a wavy line under, i.e. a̰

All the four lines drawn below are of equal length and go in the same direction and they can all be called **a**.

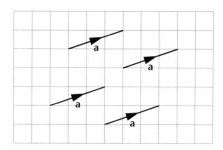

If you look at this diagram,

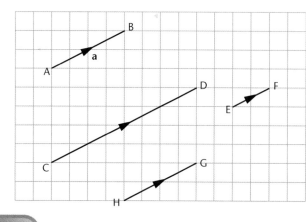

$\overrightarrow{AB}$ = **a**.

The line CD is parallel to AB and twice as long so $\overrightarrow{CD}$ = 2**a**.

EF is parallel to AB and half the length so $\overrightarrow{EF} = \frac{1}{2}$**a**.

GH is parallel and equal in length to BA (opposite direction to AB) so $\overrightarrow{GH}$ = ‾**a**.

EXAMPLE 2

For the diagram below write down the vectors $\overrightarrow{CD}$, $\overrightarrow{EF}$, $\overrightarrow{GH}$, $\overrightarrow{PQ}$. in terms of **a**.

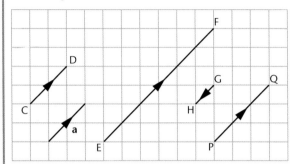

$\overrightarrow{CD} = \mathbf{a}$, $\overrightarrow{EF} = 3\mathbf{a}$, $\overrightarrow{GH} = \dfrac{^-1}{2}\mathbf{a}$, $\overrightarrow{PQ} = \dfrac{3}{2}\mathbf{a}$.

EXAMPLE 3

ABCD is a rectangle and E, F, G, H are the midpoints of the sides.

AB = **a** and AD = **b**.

Write the vectors $\overrightarrow{BC}$, $\overrightarrow{CD}$, $\overrightarrow{AE}$, $\overrightarrow{AH}$, $\overrightarrow{EG}$, $\overrightarrow{CF}$, $\overrightarrow{FH}$. in terms of **a** or **b**.

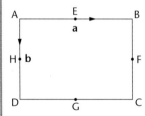

$\overrightarrow{BC} = \mathbf{b}$, $\overrightarrow{CD} = {}^-\mathbf{a}$, $\overrightarrow{AE} = \dfrac{1}{2}\mathbf{a}$, $\overrightarrow{AH} = \dfrac{1}{2}\mathbf{b}$, $\overrightarrow{EG} = \mathbf{b}$, $\overrightarrow{CF} = \dfrac{^-1}{2}\mathbf{b}$, $\overrightarrow{FH} = {}^-\mathbf{a}$.

EXERCISE 11.1A

1 Write down the column vectors for $\overrightarrow{AB}$, $\overrightarrow{CD}$, $\overrightarrow{CB}$, $\overrightarrow{AD}$ and $\overrightarrow{CA}$.

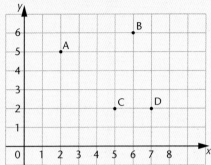

2 Write down the column vectors for $\overrightarrow{EF}$, $\overrightarrow{GH}$, $\overrightarrow{EH}$, $\overrightarrow{GF}$ and $\overrightarrow{FH}$.

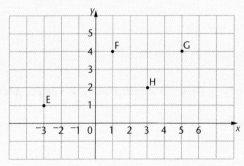

3 Find the column vector that maps:

a) (1, 2) to (1, 4)   **b)** (2, 3) to (⁻2, 3)   **c)** (1, 0) to (⁻1, 3)

d) (4, 2) to (5, 9)   **e)** (⁻3, 2) to (5, ⁻4)   **f)** (6, 1) to (0, 5)

4 For the diagram below write down the vectors $\overrightarrow{AB}$, $\overrightarrow{CD}$, $\overrightarrow{EF}$, $\overrightarrow{GH}$, $\overrightarrow{PQ}$ and $\overrightarrow{RS}$ in terms of **a**.

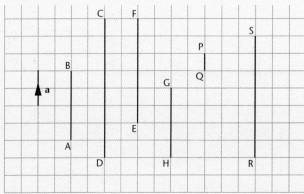

Exercise 11.1A cont'd

5 For the diagram below write down the vectors $\vec{AB}$, $\vec{CD}$, $\vec{EF}$, $\vec{GH}$, $\vec{PQ}$ and $\vec{RS}$ in terms of **a** or **b**.

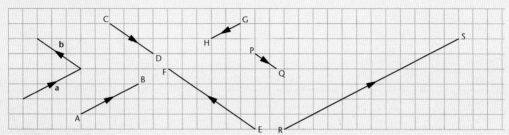

6 ABCD is a parallelogram. E, F, G, H are the mid points of the sides.

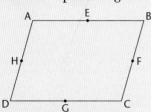

$\vec{AE} = $ **a** and $\vec{AH} = $ **b**.

Write down the vectors for $\vec{AB}$, $\vec{CD}$, $\vec{EB}$, $\vec{GD}$ $\vec{HF}$, and $\vec{FC}$ in terms of **a** or **b**.

EXERCISE 11.1B

1 Write down the column vectors for $\vec{AB}$, $\vec{CD}$, $\vec{CB}$, $\vec{AD}$ and $\vec{CA}$.

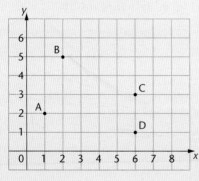

2 Write down the column vectors for $\vec{AB}$, $\vec{CD}$, $\vec{CB}$, $\vec{AD}$ and $\vec{CA}$.

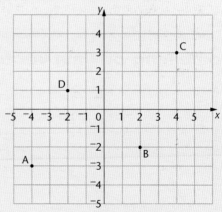

3 Find where the point is mapped by the vector.

a) (1, 2) by $\begin{pmatrix} 3 \\ 2 \end{pmatrix}$

b) (2, 3) by $\begin{pmatrix} 4 \\ 1 \end{pmatrix}$

c) (1, 0) by $\begin{pmatrix} ^-3 \\ 2 \end{pmatrix}$

d) (4, 2) by $\begin{pmatrix} 0 \\ -3 \end{pmatrix}$

e) ($^-$3, 2) by $\begin{pmatrix} ^-5 \\ ^-2 \end{pmatrix}$

f) (6, 1) by $\begin{pmatrix} ^-6 \\ ^-1 \end{pmatrix}$

Exercise 11.1B cont'd

4 For the diagram below write down the vectors $\overrightarrow{AB}$, $\overrightarrow{CD}$, $\overrightarrow{EF}$, $\overrightarrow{GH}$, $\overrightarrow{PQ}$ and $\overrightarrow{RS}$ in terms of **a**.

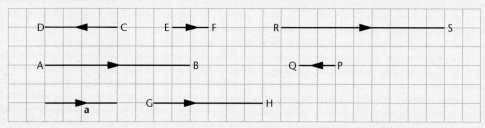

5 For the diagram below write down the vectors $\overrightarrow{AB}$, $\overrightarrow{CD}$, $\overrightarrow{EF}$, $\overrightarrow{GH}$, $\overrightarrow{PQ}$ and $\overrightarrow{RS}$ in terms of **a** or **b**.

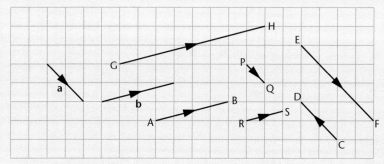

6 ABCD is a square. E, F, G, H are the midpoints of sides AB, BC, CD, DA respectively. $\overrightarrow{AB}$ = **a** and $\overrightarrow{AD}$ = **b**.

Write down the vectors $\overrightarrow{BC}$, $\overrightarrow{CD}$, $\overrightarrow{EB}$, $\overrightarrow{HD}$, $\overrightarrow{HF}$ and $\overrightarrow{FB}$ in terms of **a** or **b**.

Combining column vectors

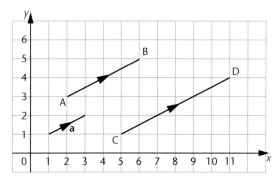

In the diagram you can see that $\overrightarrow{AB}$ = 2**a** and $\overrightarrow{CD}$ = 3**a**.

$$\mathbf{a} = \begin{pmatrix} 2 \\ 1 \end{pmatrix}$$

$$\overrightarrow{AB} = \begin{pmatrix} 4 \\ 2 \end{pmatrix} = 2 \times \begin{pmatrix} 2 \\ 1 \end{pmatrix} = 2\mathbf{a}$$

$$\overrightarrow{CD} = \begin{pmatrix} 6 \\ 3 \end{pmatrix} = 3 \times \begin{pmatrix} 2 \\ 1 \end{pmatrix} = 3\mathbf{a}$$

This shows that $p \times \begin{pmatrix} a \\ b \end{pmatrix} = \begin{pmatrix} pa \\ pb \end{pmatrix}$

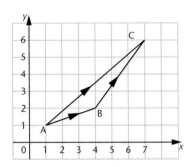

In this diagram you can see that

$\overrightarrow{AB} = \begin{pmatrix} 3 \\ 1 \end{pmatrix}$, $\overrightarrow{BC} = \begin{pmatrix} 3 \\ 4 \end{pmatrix}$, $\overrightarrow{AC} = \begin{pmatrix} 6 \\ 5 \end{pmatrix}$.

$\overrightarrow{AC} = \overrightarrow{AB} + \overrightarrow{BC} = \begin{pmatrix} 3 \\ 1 \end{pmatrix} + \begin{pmatrix} 3 \\ 4 \end{pmatrix} = \begin{pmatrix} 6 \\ 5 \end{pmatrix}$

This shows that $\begin{pmatrix} a \\ b \end{pmatrix} + \begin{pmatrix} c \\ d \end{pmatrix} = \begin{pmatrix} a+c \\ b+d \end{pmatrix}$.

Exam tip

When collecting column vectors be very careful with the signs as most errors are made in that way.

EXAMPLE 4

Given $\mathbf{a} = \begin{pmatrix} 3 \\ 1 \end{pmatrix}$, $\mathbf{b} = \begin{pmatrix} 1 \\ 3 \end{pmatrix}$, $\mathbf{c} = \begin{pmatrix} -2 \\ 1 \end{pmatrix}$

Work out the following.

a) $2\mathbf{a}$ **b)** $\mathbf{a} + 2\mathbf{b}$ **c)** $\mathbf{a} - \mathbf{b} + \mathbf{c}$ **d)** $2\mathbf{a} + \mathbf{b} - \mathbf{c}$ **e)** $\frac{1}{2}\mathbf{a}$.

a) $2\mathbf{a} = 2 \times \begin{pmatrix} 3 \\ 1 \end{pmatrix} = \begin{pmatrix} 6 \\ 2 \end{pmatrix}$ **b)** $\mathbf{a} + 2\mathbf{b} = \begin{pmatrix} 3 \\ 1 \end{pmatrix} + 2 \times \begin{pmatrix} 1 \\ 3 \end{pmatrix} = \begin{pmatrix} 5 \\ 7 \end{pmatrix}$ **c)** $\mathbf{a} - \mathbf{b} + \mathbf{c} = \begin{pmatrix} 3 \\ 1 \end{pmatrix} - \begin{pmatrix} 1 \\ 3 \end{pmatrix} + \begin{pmatrix} -2 \\ 1 \end{pmatrix} = \begin{pmatrix} 0 \\ -1 \end{pmatrix}$

d) $2\mathbf{a} + \mathbf{b} - \mathbf{c} = 2 \times \begin{pmatrix} 3 \\ 1 \end{pmatrix} + \begin{pmatrix} 1 \\ 3 \end{pmatrix} - \begin{pmatrix} -2 \\ 1 \end{pmatrix} = \begin{pmatrix} 9 \\ 4 \end{pmatrix}$ **e)** $\frac{1}{2}\mathbf{a} = \frac{1}{2} \times \begin{pmatrix} 3 \\ 1 \end{pmatrix} = \begin{pmatrix} 1\cdot5 \\ 0\cdot5 \end{pmatrix}$

EXERCISE 11.2A

1 Work out **a)** $2 \times \begin{pmatrix} 2 \\ 3 \end{pmatrix}$ **b)** $\begin{pmatrix} 6 \\ 2 \end{pmatrix} + \begin{pmatrix} 3 \\ 1 \end{pmatrix}$ **c)** $\frac{1}{2}\begin{pmatrix} 4 \\ 6 \end{pmatrix}$ **d)** $\begin{pmatrix} 3 \\ 1 \end{pmatrix} - \begin{pmatrix} 2 \\ 1 \end{pmatrix}$ **e)** $\begin{pmatrix} 3 \\ 4 \end{pmatrix} + 2 \times \begin{pmatrix} 1 \\ 4 \end{pmatrix}$.

2 Work out **a)** $2 \times \begin{pmatrix} -3 \\ 0 \end{pmatrix}$ **b)** $\begin{pmatrix} 3 \\ 1 \end{pmatrix} - \begin{pmatrix} 4 \\ 3 \end{pmatrix}$ **c)** $\frac{1}{2}\begin{pmatrix} -1 \\ -3 \end{pmatrix}$ **d)** $\begin{pmatrix} 2 \\ -1 \end{pmatrix} + 2 \times \begin{pmatrix} 2 \\ 1 \end{pmatrix}$ **e)** $\frac{1}{2}\begin{pmatrix} 1 \\ 4 \end{pmatrix} - \frac{1}{4}\begin{pmatrix} 2 \\ 4 \end{pmatrix}$.

3 Given that $\mathbf{a} = \begin{pmatrix} 6 \\ 3 \end{pmatrix}$, work out **a)** $2\mathbf{a}$ **b)** $^-\mathbf{a}$ **c)** $4\mathbf{a}$ **d)** $\frac{1}{2}\mathbf{a}$ **e)** $\frac{-1}{3}\mathbf{a}$.

4 Given that $\mathbf{a} = \begin{pmatrix} 1 \\ 3 \end{pmatrix}$, $\mathbf{b} = \begin{pmatrix} 3 \\ 4 \end{pmatrix}$, work out **a)** $3\mathbf{a}$ **b)** $\mathbf{a} + \mathbf{b}$ **c)** $\mathbf{b} - \mathbf{a}$ **d)** $2\mathbf{a} + \mathbf{b}$

     e) $3\mathbf{a} - 2\mathbf{b}$.

5 Given that $\mathbf{a} = \begin{pmatrix} 2 \\ 3 \end{pmatrix}$, $\mathbf{b} = \begin{pmatrix} -3 \\ 4 \end{pmatrix}$, $\mathbf{c} = \begin{pmatrix} -1 \\ -3 \end{pmatrix}$, work out   **a)** $3\mathbf{c}$ **b)** $4\mathbf{c} - 2\mathbf{b}$

     c) $\mathbf{a} - \mathbf{b} + \mathbf{c}$ **d)** $2\mathbf{a} + 3\mathbf{b} + 2\mathbf{c}$ **e)** $\frac{1}{2}\mathbf{a} - \mathbf{b} - \frac{1}{2}\mathbf{c}$.

EXERCISE 11.2B

1 Work out   **a)** $3 \times \begin{pmatrix} 1 \\ 4 \end{pmatrix}$ **b)** $\begin{pmatrix} 3 \\ 4 \end{pmatrix} + \begin{pmatrix} 5 \\ 8 \end{pmatrix}$ **c)** $\frac{1}{2}\begin{pmatrix} 8 \\ 10 \end{pmatrix}$ **d)** $2 \times \begin{pmatrix} 5 \\ 4 \end{pmatrix} - \begin{pmatrix} 3 \\ 4 \end{pmatrix}$

        e) $2 \times \begin{pmatrix} 1 \\ 4 \end{pmatrix} + 5 \times \begin{pmatrix} 1 \\ 2 \end{pmatrix}$.

2 Work out   **a)** $2 \times \begin{pmatrix} -1 \\ 0 \end{pmatrix}$ **b)** $\begin{pmatrix} 1 \\ 6 \end{pmatrix} - \begin{pmatrix} 7 \\ 3 \end{pmatrix}$ **c)** $\frac{1}{2}\begin{pmatrix} -2 \\ 4 \end{pmatrix}$ **d)** $\begin{pmatrix} 1 \\ -4 \end{pmatrix} - 2 \times \begin{pmatrix} 2 \\ 3 \end{pmatrix}$ **e)** $\frac{1}{2}\begin{pmatrix} 2 \\ 6 \end{pmatrix} - \frac{1}{2}\begin{pmatrix} 3 \\ -5 \end{pmatrix}$.

3 Given that $\mathbf{p} = \begin{pmatrix} 5 \\ 8 \end{pmatrix}$, work out   **a)** $4\mathbf{p}$ **b)** $^-2\mathbf{p}$ **c)** $\frac{1}{2}\mathbf{p}$ **d)** $9\mathbf{p}$ **e)** $\frac{2}{5}\mathbf{p}$.

4 Given that $\mathbf{p} = \begin{pmatrix} 4 \\ 1 \end{pmatrix}$, $\mathbf{q} = \begin{pmatrix} 5 \\ 3 \end{pmatrix}$, work out   **a)** $2\mathbf{p}$ **b)** $\mathbf{p} + \mathbf{q}$ **c)** $\mathbf{q} - \mathbf{p}$ **d)** $2\mathbf{p} + \mathbf{q}$

     e) $3\mathbf{q} - 2\mathbf{p}$.

5 Given that $\mathbf{a} = \begin{pmatrix} -2 \\ 4 \end{pmatrix}$, $\mathbf{b} = \begin{pmatrix} 3 \\ 5 \end{pmatrix}$, $\mathbf{c} = \begin{pmatrix} -2 \\ -3 \end{pmatrix}$,

   work out **a)** $3\mathbf{c}$ **b)** $3\mathbf{c} + 2\mathbf{b}$ **c)** $\mathbf{a} - \mathbf{b} + \mathbf{c}$ **d)** $\mathbf{a} + 4\mathbf{b} - 2\mathbf{c}$ **e)** $\frac{1}{2}\mathbf{a} + \mathbf{b} - \frac{1}{2}\mathbf{c}$.

Vector geometry

The resultant of two vectors

You have already seen results like $\overrightarrow{PR} = \overrightarrow{PQ} + \overrightarrow{QR}$.

In the diagram $\overrightarrow{OA} = \mathbf{a}$, $\overrightarrow{OB} = \mathbf{b}$ and OACB
is a parallelogram.

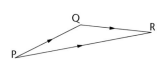

AC is parallel and equal to OA so
$\overrightarrow{AC} = \mathbf{b}$.

$\overrightarrow{OC} = \overrightarrow{OA} + \overrightarrow{AC} = \mathbf{a} + \mathbf{b}$.

$\overrightarrow{OC}$ is known as the resultant of $\mathbf{a}$ and $\mathbf{b}$.

You can also see in the diagram that $\overrightarrow{OC} = \overrightarrow{OB} + \overrightarrow{BC} = \mathbf{b} + \mathbf{a}$.

This shows that $\mathbf{a} + \mathbf{b} = \mathbf{b} + \mathbf{a}$. The vectors can be added in either order.

ACTIVITY 2

In the diagram below A, B, C, D, E are marked.

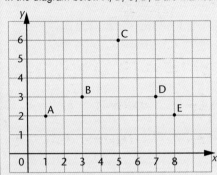

Write down five different routes from A to E. You do not have to use all the points.

For each route, add together the vectors you have used.

What do you notice?

The resultant vector is the same, no matter which way you go.

This is a very important rule which is used to find vectors in geometrical figures.

Exam tip

If you were only asked to find $\vec{AD}$ directly, it would be an example of a multi-step question and you would need to work out which other vectors to find.

EXAMPLE 5

In the triangle ABC, $\vec{AB} = \mathbf{p}$ and $\vec{AC} = \mathbf{q}$ and D is the midpoint of BC.

Work out the vectors:

a) $\vec{BC}$ **b)** $\vec{BD}$ **c)** $\vec{AD}$.

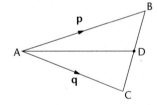

a) $\vec{BC} = \vec{BA} + \vec{AC} = {}^{-}\mathbf{p} + \mathbf{q} = \mathbf{q} - \mathbf{p}$ **b)** $\vec{BD} = \frac{1}{2}\vec{BC} = \frac{1}{2}(\mathbf{q} - \mathbf{p})$

c) $\vec{AD} = \vec{AB} + \vec{BD} = \mathbf{p} + \frac{1}{2}(\mathbf{q} - \mathbf{p}) = \mathbf{p} + \frac{1}{2}\mathbf{q} - \frac{1}{2}\mathbf{p} = \frac{1}{2}\mathbf{p} + \frac{1}{2}\mathbf{q} = \frac{1}{2}(\mathbf{p} + \mathbf{q})$.

EXAMPLE 6

In this diagram OC = 2 × OA and OD = 2 × OB.
$\vec{OA} = \mathbf{a}$ and $\vec{OB} = \mathbf{b}$.

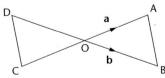

a) Work out the vectors:

(i) $\vec{OC}$ **(ii)** $\vec{OD}$

(iii) $\vec{AB}$ **(iv)** $\vec{DC}$.

b) What does this show about the lines AB and DC?

Example 6 cont'd

a) (i) $\overrightarrow{OC} = ^-2 \times \overrightarrow{OA} = ^-2\mathbf{a}$

OC is on the same line as OA, opposite direction and twice as long.

(ii) $\overrightarrow{OD} = ^-2 \times \overrightarrow{OB} = ^-2\mathbf{b}$

Same reasoning as above.

(iii) $\overrightarrow{AB} = \overrightarrow{AO} + \overrightarrow{OB} = ^-\mathbf{a} + \mathbf{b} = \mathbf{b} - \mathbf{a}$.

(iv) $\overrightarrow{DC} = \overrightarrow{DO} + \overrightarrow{OC} = 2\mathbf{b} - 2\mathbf{a} = 2(\mathbf{b} - \mathbf{a})$.

b) AB and DC are parallel and DC is twice as long as AB.

This is because the vector for DC is twice the vector for AB.

EXERCISE 11.3A

1 On a square grid with x and y from 0 to 8, plot A (1, 3) and B (3, 5).

a) Write down $\overrightarrow{AB}$ as a column vector.

b) Mark any two points as C and D and work out:

(i) $\overrightarrow{AC} + \overrightarrow{CB}$ **(ii)** $\overrightarrow{AD} + \overrightarrow{DB}$ **(iii)** $\overrightarrow{AC} + \overrightarrow{CD} + \overrightarrow{DB}$.

c) What do you notice?

2

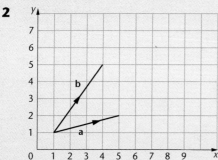

The vectors **a** and **b** are drawn on the grid. Draw the resultant of **a** and **b**.

3 A is the point ($^-2$, 1), B is the point (4, 3), C is the point (7, 4).

a) Work out the column vectors:

(i) $\overrightarrow{AB}$ **(ii)** $\overrightarrow{BC}$.

b) What can you say about A, B and C?

4

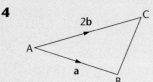

In the triangle $\overrightarrow{AB} = \mathbf{a}$, $\overrightarrow{AC} = 2\mathbf{b}$. Find the vector $\overrightarrow{BC}$ in terms of **a** and **b**.

Exercise 11.3A cont'd

5
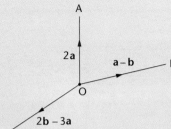

In this diagram $\overrightarrow{OA} = 2\mathbf{a}$, $\overrightarrow{OB} = \mathbf{a} - \mathbf{b}$, and $\overrightarrow{OC} = 2\mathbf{b} - 3\mathbf{a}$.

Work out as simply as possible:

a) $\overrightarrow{AB}$

b) $\overrightarrow{BC}$

c) $\overrightarrow{AC}$.

6
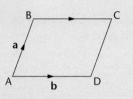

ABCD is a parallelogram. $\overrightarrow{AB} = \mathbf{a}$ and $\overrightarrow{AD} = \mathbf{b}$.

Work out the vectors $\overrightarrow{BC}$, $\overrightarrow{CD}$, $\overrightarrow{BD}$, and $\overrightarrow{AC}$ in terms of $\mathbf{a}$ and/or $\mathbf{b}$.

7
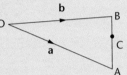

In the triangle OAB, C is a point on AB so that AC = 2 × CB. $\overrightarrow{OA} = \mathbf{a}$ and $\overrightarrow{OB} = \mathbf{b}$.

Work out the vectors $\overrightarrow{AB}$, $\overrightarrow{CB}$ and $\overrightarrow{OC}$ in terms of $\mathbf{a}$ and/or $\mathbf{b}$.

8
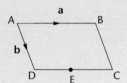

ABCD is a parallelogram. E is the midpoint of the DC line. $\overrightarrow{AB} = \mathbf{a}$ and $\overrightarrow{AD} = \mathbf{b}$.

Write down the vector $\overrightarrow{EB}$ in terms of $\mathbf{a}$ and/or $\mathbf{b}$ (multi-step).

9
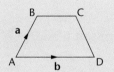

In the trapezium ABCD, AD is parallel to BC and AD = 2 × BC. $\overrightarrow{AB} = \mathbf{a}$ and $\overrightarrow{AD} = \mathbf{b}$.

Write down the vector $\overrightarrow{CD}$ in terms of $\mathbf{a}$ and $\mathbf{b}$ (multi-step).

10
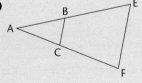

Triangle AEF is a 3 × enlargement of triangle ABC. $\overrightarrow{AB} = \mathbf{a}$ and $\overrightarrow{AC} = \mathbf{b}$.

a) Write down the vectors $\overrightarrow{AE}$, $\overrightarrow{AF}$, $\overrightarrow{BC}$, $\overrightarrow{EF}$ in terms of $\mathbf{a}$ and/or $\mathbf{b}$.

b) What do the vectors show about BC and EF?

1 On a square grid with x and y from 0 to 8, plot A (2, 3) and B (3, 1).

a) Write down $\overrightarrow{AB}$ as a column vector.

b) Mark any two points as C and D and work out

(i) $\overrightarrow{AC} + \overrightarrow{CB}$ **(ii)** $\overrightarrow{AD} + \overrightarrow{DB}$ **(iii)** $\overrightarrow{AC} + \overrightarrow{CD} + \overrightarrow{DB}$.

c) What do you notice?

2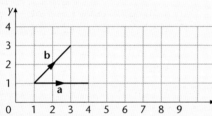
The vectors **a** and **b** are drawn on the grid.

Draw the resultant of 2**a** and **b**.

3 A is the point (2, 1), B is the point (4, 4), C is the point (7, 4), D is the point (3, ⁻2).

a) Work out the column vectors **(i)** $\overrightarrow{AB}$ **(ii)** $\overrightarrow{CD}$.

b) What can you say about AB and CD?

4 In the triangle ABC, $\overrightarrow{AB} = 2\mathbf{a}$ and $\overrightarrow{CB} = 3\mathbf{b}$.
Work out the vector $\overrightarrow{AC}$.
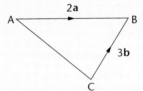

5 In this diagram $\overrightarrow{OA} = 2\mathbf{a}$, $\overrightarrow{OB} = 2\mathbf{a} + 3\mathbf{b}$ and $\overrightarrow{OC} = 3\mathbf{b}$.
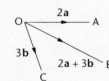

a) Write down the vectors $\overrightarrow{AB}$ and $\overrightarrow{BC}$ in terms of **a** and/or **b**.

b) What can you say about the shape OABC?

6 Work out the vector $\overrightarrow{AB}$ for this shape.
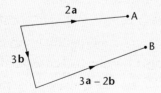

Exercise 11.3B cont'd

7 ABCD is a parallelogram. E, F, G, H are the midpoints of the sides.
$\overrightarrow{AE} = \mathbf{a}$ and $\overrightarrow{AH} = \mathbf{b}$.

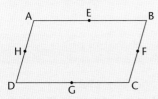

a) Write down the vectors $\overrightarrow{EB}$, $\overrightarrow{BF}$, $\overrightarrow{EF}$, $\overrightarrow{HD}$, $\overrightarrow{DG}$ and $\overrightarrow{HG}$ in terms of **a** and/or **b**.

b) What does this show about EF and HG?

8

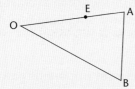

OAB is a triangle with C a point on OA so that OE = 2 × EA.

$\overrightarrow{OA} = \mathbf{a}$ and $\overrightarrow{OB} = \mathbf{b}$.

Work out the vector $\overrightarrow{EB}$ in terms of **a** and/or **b** (multi-step).

9

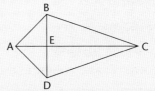

ABCD is a kite. E is the point where the diagonals cross, BE = ED and CE = 3 × AE in length. $\overrightarrow{AB} = \mathbf{a}$ and $\overrightarrow{AD} = \mathbf{b}$.

a) Work out the vectors $\overrightarrow{BD}$, $\overrightarrow{BE}$, $\overrightarrow{AE}$, $\overrightarrow{EC}$ and $\overrightarrow{BC}$ in terms of **a** and/or **b**.

b) Explain why the vectors show that BC is not parallel to AD.

10 In the triangle OCD, AC = 3 × OA and BD = 3 × OB.
$\overrightarrow{OA} = \mathbf{a}$ and $\overrightarrow{OB} = \mathbf{b}$.

a) Use vectors to show that AB is parallel to CD.

b) What is the ratio of the lengths of AB and CD?

Key ideas

- A vector has magnitude (length) and direction but can start at any point.
- Lines that are parallel have vectors that are multiples of each other.
- If $\overrightarrow{BC} = n \times \overrightarrow{AB}$, then ABC is a straight line and BC is n times AB in length.
- If $\overrightarrow{CD} = n \times \overrightarrow{AB}$, then AB and CD are parallel and CD is n times AB in length.
- To add or subtract column vectors, add or subtract the two components separately.
- To multiply a column vector by a number, multiply each component by that number.
- The resultant of two vectors is the third side of the triangle formed by those vectors.
- The vector $\overrightarrow{AB}$, is equal to the sum of the vectors $\overrightarrow{AC}$, + $\overrightarrow{CD}$ + ··· + $\overrightarrow{PQ}$ + $\overrightarrow{QB}$, where C, D, E, ..., P, Q are any points.

Comparing sets of data

You should already know

- how to plot and interpret cumulative frequency graphs and box plots
- how to plot and interpret histograms
- how to find means, medians, modes
- how to find a range, quartiles and interquartile range.

Often, the purpose of calculating some statistics for a distribution is to be able to compare the distribution with others. Are the people in this group taller or shorter than average? Do people shopping at this centre on Saturdays spend more than those shopping on Tuesdays? How does the life of these light-bulbs compare with the previous design produced by this company?

The valid interpretation of the statistics you have calculated is the most important part of any statistics project. This chapter gives you more practice in these skills.

ACTIVITY 1

Work in small groups.

Look at, or remember, the GCSE statistic project that you carried out, if you have done it already.

Make a list of some of the conclusions you came to in that project.

or

Choose one of these ideas:

1 In your year group, are the boys taller than the girls?

2 Are people's speeds of reaction determined by gender or age?

Make notes on what data you would need to collect and how you would collect it. What would you be looking for to make a conclusion to your chosen problem?

When comparing sets of data you usually need to compare two types of statistics:

- averages
- spread.

When comparing, you need to compare the same types of data in both distributions. For instance, comparing the mean in one with the median in the other will not tell you anything helpful.

Exam tip

Make sure any comparisons you make are related to the context.

EXAMPLE

These two box plots show the results for the life of two different types of torch battery. Which type of battery would you choose, and why?

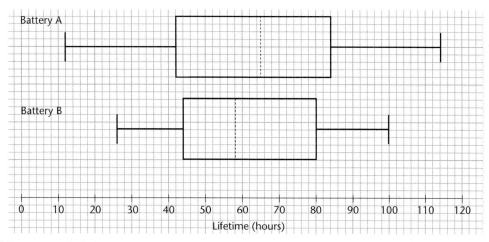

Possible responses include:

- I would choose Battery A since it has a greater median, showing it lasts longer on average.
- Battery B – although its average lifetime is less, its smaller range shows its performance is more consistent.
- Battery B – I need to be able to rely on the battery lasting a good length of time – one battery of type A lasted only 12 hours.

When comparing two distributions, ideas of skewness may help you, although they are not formally needed in the GCSE programme of study.

Look at the shapes of these three histograms.

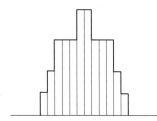

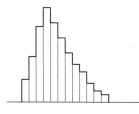

This is a symmetric distribution.

mean = mean = mode

In this distribution, the longer tail lies to the right.

This is called a positive skew.

mode < median < mean

In this distribution, the longer tail lies to the left.

This is called a negative skew.

mean < median < mean

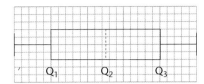

Skewness may also be identified form the median and quartiles, for instance using box plots.

This is a symmetric distribution.

$$Q_3 - Q_2 = Q_2 - Q_1$$

This has a positive skew. The gap in the box to the right of the median is larger than the gap to the left.

$$Q_3 - Q_2 > Q_2 - Q_1$$

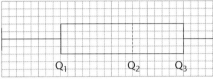

This has a negative skew. The gap in the box to the right of the median is smaller than the gap to the left.

$$Q_3 - Q_2 < Q_2 - Q_1$$

Sometimes, graphs in newspaper articles or in advertisements are quite complicated. You need to be able to look for the main features of what the graphs tell you. You may sometimes find that claims are made which are not backed up by the graphs!

Exam tip

Look at the title of a graph and at the labels on the axes. Then look at the information shown by the plotted points and any trend lines.

ACTIVITY 2

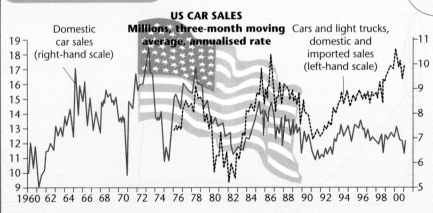

1 Look at the graph and write down the answers to these questions.

 a) Describe fully what the peak of the red graph shows.

 b) What was the lowest value for the 'cars and light trucks' graph during the 1990s?

 c) Criticise two aspects of the presentation of this graph.

2 Working in pairs or in a group, write more questions about this graph, and answer each other's questions.

Chapter 12 *Comparing sets of data*

ACTIVITY 3

Collect graphs from newspapers and magazines. Look in particular for graphs which are designed to make comparisons.

Work in groups, discussing how much information is given on the graphs. For instance, can you interpret them without needing to refer to the accompanying articles?

Say also what could be done to improve the presentation and usefulness of the graphs.

You may like to make a poster of your findings.

EXERCISE 12.1A

1 The masses of 400 potatoes of each of two varieties were measured.

Here are the results.

a) Draw a cumulative graph, showing both of these distributions on the same diagram.

b) Compare the distributions.

Mass (m g)	Frequency for variety A	Frequency for variety B
$50 < m \le 100$	0	28
$100 < m \le 150$	43	65
$150 < m \le 200$	88	96
$200 < m \le 250$	137	89
$250 < m \le 300$	79	75
$300 < m \le 350$	53	47

2 The masses of some tomatoes of variety A were found to be:

a) Draw a cumulative frequency diagram, with a summary box diagram below it, to represent this distribution.

Another sample of tomatoes of variety B had this summary data.

Mass (m g)	Frequency
$0 < m \le 20$	3
$20 < m \le 30$	6
$30 < m \le 40$	11
$40 < m \le 50$	18
$50 < m \le 60$	10
$60 < m \le 70$	2

Number in sample	30	Median	46 g
Lower quartile	28 g	Upper quartile	53 g

b) Make three comparisons between these distributions.

3 These box plots show the noon temperatures for one month in Guildford and Torquay.

a) Which town was generally hotter? Show how you decide.

b) Comment on the variability of temperature in the two towns.

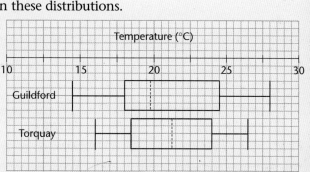

Chapter 12 *Comparing sets of data*

Exercise 12.1A cont'd

4 This histogram shows the population living in Aveford.

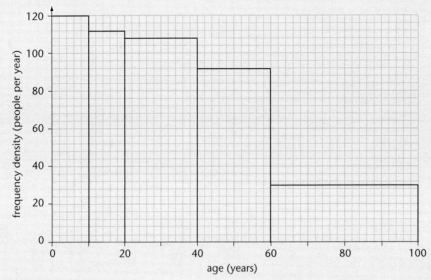

a) How many people between the ages of 20 and 40 live in Aveford?
The table below shows the population living in Banton.

Age (*a* years)	$0 \leq a < 10$	$10 \leq a < 20$	$20 \leq a < 40$	$40 \leq a < 60$	$60 \leq a < 100$
Frequency	760	920	1680	2040	1200

b) Construct a histogram to illustrate these data.

c) Make two comparisons between the populations of Aveford and Banton.

5 These two histograms show
the ages of passengers on a
plane to the Bahamas and
on a plane to Majorca.

a) Comment on the relative
shape of the histograms.

b) Construct frequency
tables and calculate the
mean age of the
passengers on each plane.

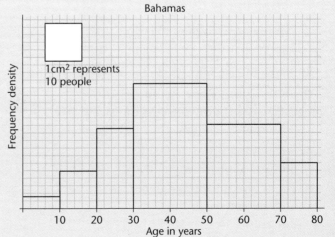

Exercise 12.1A cont'd

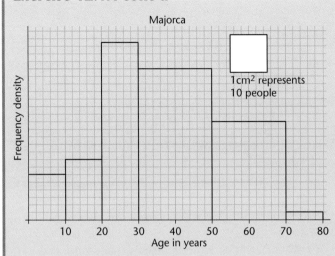

Majorca

1cm² represents 10 people

6   The graphs show the results of 700 candidates in their English and Maths examinations.

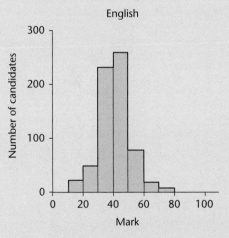

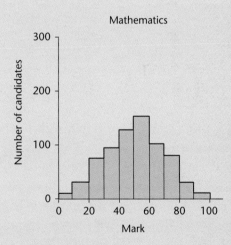

Comment on what the relative shapes of the histograms tell you.

Exercise 12.1A cont'd

7 To compare the length of different types of leaves, some lilac and lime leaves were measured correct to the nearest millimetre.

Length (mm)	30–39	40–49	50–59	60–69	70–79	80–89	90–99	100–109	110–119
Frequency [lilac]	0	3	3	4	7	11	12	7	3
Frequency [lime]	0	2	3	4	13	12	12	4	0

Compare these distributions, making appropriate calculations and/or drawing appropriate graphs to enable you to do so.

EXERCISE 12.1B

1 Here are the lengths of a sample of two varieties of runner bean, Longerpod and Red Queen.

Length L cm	Frequency for Longerpod	Frequency for Red Queen
$10 < L \leq 15$	0	3
$15 < L \leq 20$	14	12
$20 < L \leq 25$	27	18
$25 < L \leq 30$	17	22
$30 < L \leq 35$	2	5

a) Draw a cumulative graph, showing both of these distributions on the same diagram.

b) Compare the distributions.

2 Mr Banks decided to check on the lengths of phone calls made by his son Simon and daughter Joanne. He kept a check for 40 of each of their calls. The graph shows the cumulative frequency of the lengths of each of their calls.

Use the median and interquartile range to compare the length of their calls.

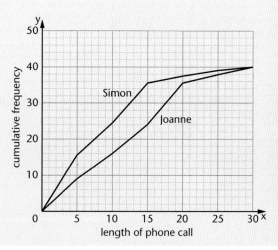

Exercise 12.1B cont'd

3

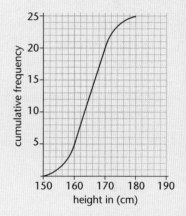

The cumulative frequency graph shows the spending money that two classes, A and B, of children have each week.

a) What information is given by the point where the two graphs cross each other?

b) Make two other comparisons between these two distributions.

4 The cumulative frequency graph shows the heights of a class of children.

a) If, a year later, they have all grown by 4 cm, describe the shape of the cumulative frequency graph then, compared with now.

b) If, instead, the shorter children grow by 3 cm but the tallest children grow by 6 cm, sketch the graphs of then and now on the same set of axes, clearly showing the comparison.

5 The heights of students at Standish School were measured to the nearest centimetre.

Height (h cm)	Frequency
$141 \leq h \leq 150$	8
$151 \leq h \leq 155$	22
$156 \leq h \leq 160$	21
$161 \leq h \leq 165$	16
$166 \leq h \leq 170$	9
$171 \leq h \leq 175$	8
$176 \leq h \leq 190$	3

The histogram overleaf shows the heights of students in one year group.

This table shows the heights of another year group.

Compare the mean heights of the two groups.

Chapter 12 *Comparing sets of data*

Exercise 12.1B cont'd

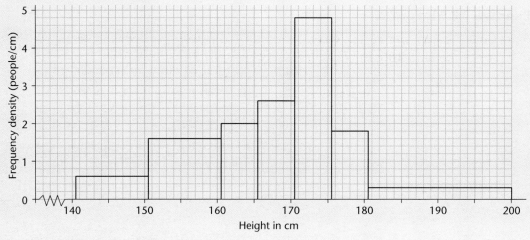

6   These two histograms show
     the age distribution of the
     Wellfit and Superhealth
     fitness clubs.

     a) How many members does
     each of these clubs have?

     b) Make two other
     comparisons between
     the membership of the
     two clubs.

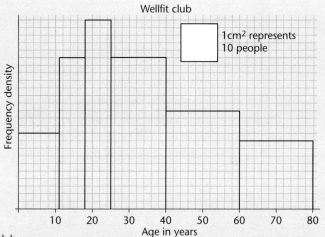

Wellfit club

1cm² represents 10 people

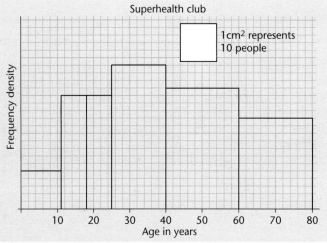

Superhealth club

1cm² represents 10 people

Exercise 12.1B cont'd

7 During the foot and mouth epidemic of 2001, a newspaper published these graphs on 9 April, comparing it with the outbreak of the disease in 1967.

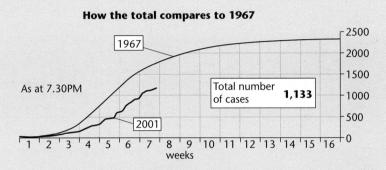

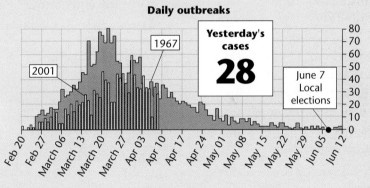

a) Describe the 1967 epidemic, as shown by these graphs.

b) Compare the 2001 outbreak with that of 1967.

Key ideas

When comparing sets of data:

● **compare averages and compare spreads**

● **compare the same type of data in each distribution**

● **use calculations and/or statistical diagrams**

● **relate your comparisons to the context of the data.**

Chapter 12 *Comparing sets of data*

Revision exercise

1 O is the centre of the circle. Calculate the area of the shaded segment.

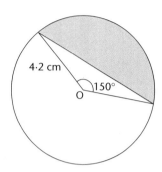

4·2 cm

150°

O

2 A solid cone and a solid cylinder both have base radius 6 cm. The height of the cylinder is 4 cm. The cone and the cylinder each have the same volume.

a) Find the height of the cone.

b) Calculate the curved surface area of the cylinder.

3 A sphere has volume 50 cm³. Calculate its surface area.

4

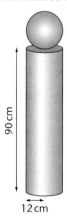

90 cm

12 cm

A traffic bollard consists of a sphere on top of a cylinder. The radii of the sphere and cylinder are each 12 cm. The height of the cylinder is 90 cm.

One litre of black paint covers 4 m². How many of these bollards can be painted with 10 litres of paint?

5 Simplify

a) $\dfrac{x}{2} + \dfrac{x+2}{3}$

b) $\dfrac{2x-1}{4} - \dfrac{2x+3}{5}$

c) $\dfrac{1}{x+1} + \dfrac{2}{x-2}$

d) $\dfrac{2x}{x-1} - \dfrac{x-1}{x+2}$

e) $\dfrac{3x^2 + 9x}{x^2 + 4x + 3}$

6 Solve the equations.

a) $x(x-2) - 2x(x-3) = 12 - x^2$

b) $\dfrac{2x}{3} + \dfrac{x-2}{2} = 1$

c) $x + 1 = \dfrac{16}{x+1}$

d) $\dfrac{x^2}{3} - \dfrac{x}{3} - 4 = 0$

e) $\dfrac{1}{x+1} = \dfrac{4}{3x+2}$

f) $x + 2 = \dfrac{15}{x}$

g) $\dfrac{5}{x+1} - \dfrac{2}{x-1} = \dfrac{1}{3}$

7 Given that $\mathbf{a} = \begin{pmatrix} 1 \\ 2 \end{pmatrix}$, $\mathbf{b} = \begin{pmatrix} {}^-2 \\ 1 \end{pmatrix}$, $\mathbf{c} = \begin{pmatrix} {}^-1 \\ {}^-3 \end{pmatrix}$, work out:

a) $2\mathbf{a}$ **b)** $\mathbf{a} - \mathbf{b}$

c) $\mathbf{a} - \mathbf{b} + \mathbf{c}$ **d)** $\mathbf{a} + 2\mathbf{b}$

e) $3\mathbf{a} + 2\mathbf{c}$ **f)** $\dfrac{1}{2}\mathbf{a}$

g) $2\mathbf{a} - 3\mathbf{c}$ **h)** $\dfrac{1}{2}\mathbf{b} - \dfrac{1}{2}\mathbf{c}$

i) $\mathbf{a} - \dfrac{1}{2}\mathbf{c} - \mathbf{b}$

8 What are the coordinates of the image when

a) the point ($^-2$, 1) is translated by $\begin{pmatrix} 1 \\ 2 \end{pmatrix}$

b) the point (4, 3) is translated by $\begin{pmatrix} ^-4 \\ ^-3 \end{pmatrix}$

c) the point (2, $^-4$) is translated by $\begin{pmatrix} 3 \\ ^-1 \end{pmatrix}$?

9 In the triangle ABC, $\overrightarrow{AB} = \mathbf{a}$ and $\overrightarrow{AC} = 2\mathbf{b}$.

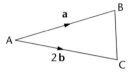

Write down the vector $\overrightarrow{BC}$ in terms of **a** and **b**.

10 In the diagram $\overrightarrow{OA} = \mathbf{a}$, $\overrightarrow{OB} = 2\mathbf{b} - \mathbf{a}$, $\overrightarrow{OC} = 6\mathbf{b} - 5\mathbf{a}$

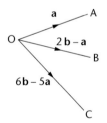

a) Work out the vectors $\overrightarrow{AB}$ and $\overrightarrow{BC}$ in terms of **a** and/or **b**.

b) What can you say about AB and BC?

11 ABCD is a parallelogram. E, F, G and H are the midpoints of the sides.

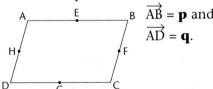

$\overrightarrow{AB} = \mathbf{p}$ and $\overrightarrow{AD} = \mathbf{q}$.

a) Find the vectors $\overrightarrow{EB}$, $\overrightarrow{BF}$, $\overrightarrow{EF}$, $\overrightarrow{HD}$, $\overrightarrow{DG}$ and $\overrightarrow{HG}$ in terms of **p** and **q**.

b) What can you say about HG and EF?

12 ABCD is a rectangle. E is a point on the diagonal AC so that AE = 2 × EC. $\overrightarrow{AB} = \mathbf{p}$ and $\overrightarrow{AD} = \mathbf{q}$.

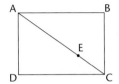

Work out the vector $\overrightarrow{EB}$ (multi-step).

13 ABCD is a quadrilateral with $\overrightarrow{AB} = 3\mathbf{p}$, $\overrightarrow{AD} = \mathbf{q}$, and $\overrightarrow{BC} = \mathbf{q} + 2\mathbf{p}$.

Use vectors to identify the type of quadrilateral.

14 The table shows the prices of a sample of 100 houses in the North West of England.

Price (£000)	Number of houses
$20 < x \le 40$	4
$40 < x \le 60$	15
$60 < x \le 80$	27
$80 < x \le 100$	41
$100 < x \le 120$	10
$120 < x \le 140$	3

Find the median and interquartile range for this sample. A similar sample in the South East gave a median of £112 000 and an interquartile range of £150 000. Compare the two areas.

15 These cumulative frequency diagrams show the marks obtained in examinations in French and English by 200 pupils in Year 8.

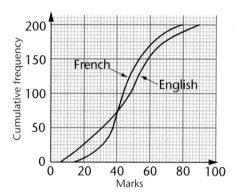

a) Draw box plots for each of the languages.

b) Use the median and interquartile range for each subject to compare the results.

16 The heights of students in two classes are measured. The results are given in the table below.

Class 11A	
Height (H cm)	Frequency
$130 \leq H < 140$	1
$140 \leq H < 150$	4
$150 \leq H < 160$	9
$160 \leq H < 170$	8
$170 \leq H < 180$	2
$180 \leq H < 190$	2

Class 11B	
Height (H cm)	Frequency
$120 \leq H < 130$	4
$130 \leq H < 140$	5
$140 \leq H < 150$	8
$150 \leq H < 160$	3
$160 \leq H < 170$	3
$170 \leq H < 180$	1

a) Show the data on two histograms.

b) Compare the heights of the students in the two classes.

13 Further simultaneous equations

You should already know

- how to solve linear simultaneous equations algebraically
- how to solve quadratic equations algebraically
- how to solve quadratic equations graphically
- that the equation of a circle radius r, with centre at the origin, is $x^2 + y^2 = r^2$.

ACTIVITY

Remind yourself about simultaneous equations.

Try to solve these using an algebraic method.

a) $x - 3y = 1$
$2x + y = 9$

b) $2x - 3y = 0$
$3x + 4y = 17$

c) $7x + 2y = 11$
$3x - 5y = {}^-7$

d) $5x + 3y = 9$
$3x - 2y = 13$

e) $4x + 3y = 5$
$6x + 7y = 10$ ○

Further simultaneous equations

The method that was covered in an earlier module was the method of elimination.

You most likely used this as the method in Activity 1.

Exam tip

Always check the simultaneous equation answers. If there are any errors, check the adding or subtracting as this is where most errors occur.

EXAMPLE 1

Solve the simultaneous equations:

$2x + 3y = 13$, $6x + 2y = 11$

$2x + 3y = 13$ [1]

$6x + 2y = 11$ [2]

a) To solve by elimination the easiest way is to multiply [1] by 3 and then subtract.

$[1] \times 3$ $6x + 9y = 39$ [3]

           $6x + 2y = 11$ [2]

$[3] - [2]$ $7y = 28$

              $y = 4.$

Substitute in [1] $2x + 12 = 13$

                $2x = 1$

                $x = \dfrac{1}{2}$

Check in [2] $6x + 2y = 3 + 8 = 11$, which is correct

Solution: $x = \dfrac{1}{2}, y = 4.$

These equations can also be solved by substitution.

b) $2x + 3y = 13$ [1]

               $6x + 2y = 11$ [2]

First arrange [1] to make x the subject.

            $2x = 13 - 3y$

             $x = \dfrac{13 - 3y}{2}$

Substitute this in [2]

       $6\left(\dfrac{13 - 3y}{2}\right) + 2y = 11$

          $39 - 9y + 2y = 11$

Cancelling 6 and 2 and expanding the bracket.

            $39 - 7y = 11$

              $7y = 28$

               $y = 4.$

The rest is the same as in a).

Before looking at the harder simultaneous equations it is best to see another of the type you have already met, solved by substitution.

EXAMPLE 2

Solve by substitution the simultaneous equations:

          $3x + 2y = 12$ [1]

and

          $5x - 3y = 1$ [2]

Either x or y can be substituted.

Here it is easiest is to substitute for y from [1] into [2]

[1] gives $2y = 12 - 3x$

           $y = \dfrac{12 - 3x}{2}$

Substitute in [2]

   $5x - 3\left(\dfrac{12 - 3x}{2}\right) = 1$ Multiply through by 2.

       $10x - 3(12 - 3x) = 2$

       $10x - 36 + 9x = 2$

            $19x = 38$

             $x = 2.$

Substitute in [1]

         $6 + 2y = 12$

            $2y = 6$

             $y = 3.$

Check in [2]. $5x - 3y = 10 - 9 = 1.$

Answers $x = 2, y = 3.$

One harder type of simultaneous equations to solve includes one linear equation and one quadratic, e.g. $y = 3x + 2$ and $y = x^2 - 2x + 8$. These have previously been solved by a graphical method. To solve these algebraically, the substitution method needs to be used.

EXAMPLE 3

Solve the simultaneous equations:

$$y = x^2 + 3x - 7 \qquad [1]$$

and

$$y = x - 4 \qquad [2].$$

a) graphically using values of x from $^-5$ to $+2$

b) algebraically.

a) $y = x^2 + 3x - 7$

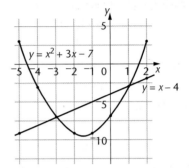

x	$^-5$	$^-4$	$^-3$	$^-2$	$^-1$	0	1	2
x^2	25	16	9	4	1	0	1	4
$+ 3x$	$^-15$	$^-12$	$^-9$	$^-6$	$^-3$	0	3	6
$^-7$	$^-7$	$^-7$	$^-7$	$^-7$	$^-7$	$^-7$	$^-7$	$^-7$
$y = x^2 + 3x - 7$	3	$^-3$	$^-7$	$^-9$	$^-9$	$^-7$	$^-3$	3

$y = x - 4$

x	$^-5$	0	2
y	$^-9$	$^-4$	$^-2$

The curve and the line cross at $x = 1$, $y = ^-3$ and $x = ^-3$, $y = ^-7$.

b) Substitute y from [2] into [1].

Substituting gives $\qquad x - 4 = x^2 + 3x - 7.$

$$x^2 + 2x - 3 = 0 \qquad \text{rearranging}$$

$$(x - 1)(x + 3) = 0$$

$$x \qquad\qquad = 1 \text{ or } ^-3.$$

Substitute in [1] for $x = 1$, $y = 1 + 3 - 7 = ^-3$

for $x = ^-3$, $y = 9 - 9 - 7 = ^-7$

Check in [2] for $x = 1$, $y = ^-3$, $^-3 = 1 - 4 = ^-3$ $\qquad$ checks

for $x = ^-3$, $y = ^-7$, $^-7 = 3 - 4 = ^-7$ $\qquad$ checks

The solutions are $x = 1$, $y = ^-3$ or $x = ^-3$, $y = ^-7$.

EXAMPLE 4

Solve, algebraically, these simultaneous equations:

$$3x + 2y = 7 \text{ [1]}$$

and $$y = x^2 - 2x + 3 \text{ [2]}$$

Either x or y can be substituted but the easiest way is to substitute for y from [2] into [1]

$$3x + 2(x^2 - 2x + 3) = 7$$
$$3x + 2x^2 \, ^-4x + 6 = 7$$
$$2x^2 - x - 1 = 0$$
$$(2x + 1)(x - 1) = 0$$
$$x = -\frac{1}{2} \text{ or } 1$$

> **Exam tip**
>
> Always substitute for the letter that needs the least manipulation, normally the letter y.

Substitute in [1] for $x = -\frac{1}{2}, -\frac{2}{3} + 2y = 7, 2y = 8\frac{1}{2}, y = 4\frac{1}{4}$

for $x = 1, 3 + 2y = 7, 2y = 4, y = 2.$

Check in [2] for $x = -\frac{1}{2}, y = 4\frac{1}{4}, y = \left(-\frac{1}{2}\right)^2 - 2\left(-\frac{1}{2}\right) + 3 = \frac{1}{4} + 1 + 3 = 4\frac{1}{4}$

for $x = 1, y = 2, y = 4 - 4 + 3 = 3$ both check.

Solutions: $x = -\frac{1}{2}, y = 4\frac{1}{4},$ or $x = 1, y = 2.$

EXERCISE 13.1A

Solve the following simultaneous equations by the method of substitution.

1 $\quad y = 2x - 1$
$\quad x + 2y = 8$

2 $\quad 3y = 11 - x$
$\quad 3x - y = 3$

3 $\quad 3x + 2y = 7$
$\quad 2x - 3y = \, ^-4$

4 $\quad 3x - 2y = 3$
$\quad 2x - y = 4$

5 $\quad y = 10 - 2x$
$\quad y = x^2 - 5x + 6$

6 $\quad y - 4x - 7 = 0$
$\quad y = x^2 - 3x - 1$

7 $\quad 3x + 2y = 7$
$\quad y = x^2 - x + 3$

8 $\quad x + 2y = 8$
$\quad y = x^2 + x + 3$

9 Solve algebraically the equations
$y = x^2 - 5x + 5$ and $2x + y = 9.$

10 Solve algebraically the equations
$y = x^2 - 3x - 1$ and $4x + y = 5.$

EXERCISE 13.1B

Solve the following simultaneous equations by the method of substitution.

1 $y = 2x - 3$
    $7x - 4y = 10$

2 $4x - 2y = 3$
    $x - y = 0$

3 $3x - y = 7$
    $5x + 2y = 8$

4 $2x + 3y = 7$
    $5y = 11 - 3x$

5 $y = 4 - 3x$
    $y = x^2 - 6x - 6$

6 $y = 2x - 3$
    $y = x^2 - 4x + 5$

7 $y = x^2 + x + 3$
    $2x + y = 1$

8 $y = x^2 + x - 2$
    $x + 5y + 2 = 0$

9 Solve algebraically the equations
    $y = x^2 + 3$ and $y = 3x + 7$.

10 Solve algebraically the equations
    $y = x^2 - 5x + 3$ and $7x + 2y = 11$.

So far in this chapter you have solved equations where, graphically, a straight line intersects a parabola.

In Stage 9 you learnt how to solve equations where, graphically, a straight line intersects a circle with centre at the origin. Simultaneous equations of this form can also be solved by substitution.

Exam tip

Reminder: The equation of a circle centre (0,0) and radius r is $x^2 + y^2 = r^2$.

EXAMPLE 5

Use algebra to solve simultaneously the equations
    $x^2 + y^2 = 25$ [1]
        $y = x + 1$ [2].
Substitute [2] in [1]
    $x^2 + (x + 1)^2 = 25$.
    $x^2 + x^2 + 2x + 1 = 25$ **Expand the bracket**
    $2x^2 + 2x - 24 = 0$ **Collect the terms**
    $x^2 + x - 12 = 0$ **Divide by 2**
    $(x + 4)(x - 3) = 0$
        $x = {}^-4$ or $+ 3$

Substitute in [2] for $x = {}^-4$, $y = {}^-4 + 1 = {}^-3$
for $x = 3$, $y = 3 + 1 = 4$.
Check in [1] for
for $x = {}^-4$, $y = {}^-3$; $x^2 + y^2 = 16 + 9 = 25$
for $x = 3$, $y = 4$; $x^2 + y^2 = 9 + 16 = 25$ checks.
The solutions are
        $x = {}^-4$, $y = 3$ or $x = 3$, $y = 4$.

EXERCISE 13.2A

Use algebra to solve simultaneously the following equations.

1 $x^2 + y^2 = 49$ and $y = 7 - x$.

2 $x^2 + y^2 = 169$ and $y = x + 7$.

3 $x^2 + y^2 = 25$ and $x + y = 5$.

4 $x^2 + y^2 = 100$ and $y = x + 2$.

5 $x^2 + y^2 = 64$ and $y = 2x + 8$.

Use algebra to solve the following simultaneous equations, giving the solutions correct to 2 d.p.

6 $x^2 + y^2 = 4$ and $y = x$.

7 $x^2 + y^2 = 5$ and $y = x + 2$.

8 $x^2 + y^2 = 36$ and $y = 2x - 1$.

EXERCISE 13.2B

Use algebra to solve simultaneously the following equations

1 $x^2 + y = 4$ and $y = 2 - x$.   **2** $x^2 + y^2 = 225$ and $y = x + 3$.

3 $x^2 + y^2 = 9$ and $x + y = 3$.   **4** $x^2 + y^2 = 100$ and $y = 14 - x$.

5 $x^2 + y^2 = 34$ and $y = x - 2$.

Use algebra to solve the following simultaneous equations, giving the solutions correct to 2 d.p.

6 $x^2 + y^2 = 16$ and $y = {}^-x$.   **7** $x^2 + y^2 = 10$ and $y = x + 3$.

8 $x^2 + y^2 = 25$ and $y = 3x + 1$.

Key ideas

- When solving two linear simultaneous equation algebraically, the methods of elimination or substitution can be used.

- When solving, algebraically, simultaneous equations consisting of a linear equation and a quadratic or circle equation, use substitution. Use the linear equation to find one letter in terms of the other, then substitute this in the remaining equation.

14 *Trigonometrical functions*

You should already know

- **how to use your calculator with trigonometrical functions**
- **Pythagoras' theorem.**

ACTIVITY 1

You have used sin, cos and tan in right-angled triangles, so the angles have been acute. You have also used sin and cos with obtuse angles when using the sine and cosine rules.

Enter some non-acute angles in your calculator, e.g. $^-40°$, $120°$, $270°$, $300°$.

Find the sine of these angles. Then find the inverse sine of your answer. What do you notice about your answers for sine?

What do you notice about your answers for inverse sine?

Repeat with some other angles, until you think you know what is happening.

Have you got enough values to sketch or draw accurately a graph of $y = \sin x°$? Or you could use a graphics calculator or graph-drawing program to draw it.

Similarly, try this activity with cos and tan.

Trigonometrical functions of any angle

You have already learnt how to use the sine, cosine and tangent functions with angles up to 90°. With the sine and cosine rules, you have used sin and cos for angles up to 180°. However, if you enter any angle on your calculator, you will find it will give you a value for the sin, cos or tan of that angle.

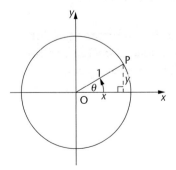

In this diagram, for an acute angle you can see that

$\cos\theta = \dfrac{x}{1}$ so $x = \cos\theta$

$\sin\theta = \dfrac{y}{1}$ so $y = \sin\theta$

$\tan\theta = \dfrac{y}{x} = \dfrac{\sin\theta}{\cos\theta}$

So P has coordinates $(\cos\theta, \sin\theta)$.

For other angles, the trigonometric functions are defined in a similar way, where the angle is measured anticlockwise from the x-axis.

241

In this diagram you can see that cos 210° and sin 210° are both negative.

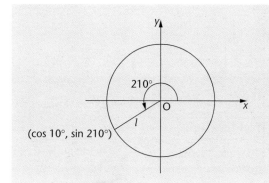

You can use symmetry to see that

cos 210° = ⁻cos 30°

sin 210° = ⁻sin 30°

$\tan 210° = \dfrac{\sin 210°}{\cos 210°} = \dfrac{\sin 30°}{\cos 30°} = \tan 30°$

so tan 210° is positive.

Continuing in a similar way, we obtain these graphs.

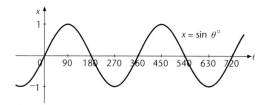

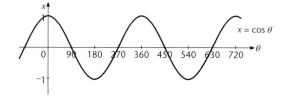

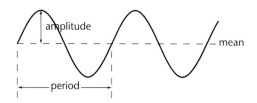

These graphs are both wave-shaped and repeat every 360°. The length of a repeating pattern is called the period. Here the period is 360°.

For a wave, the amount it varies from its mean is called the amplitude. For these graphs the amplitude is 1.

The graph of $y = \tan\theta$ has a different shape. Since $\tan\theta = \dfrac{\sin\theta}{\cos\theta}$ there is a problem when $\cos\theta = 0$. For instance, if you try using your calculator to find tan 90°, you will get an error message. Try entering different angles from 80°, getting closer to 90°, and you will see why.

The graph of $x = \cos\theta$ shows that $\cos\theta = 0$ when $\theta = 90°, 270°, 450°$, etc., and the graph of $y = \tan\theta$ is discontinuous at these values.

The graph of $y = \tan\theta$ is not a wave but does repeat. Its period is 180°. The graph approaches, but never meets, the line $\theta = 90°$, $\theta = 270°$, etc. These lines are called **asymptotes** and are shown on the graph by dotted lines.

You may need to draw accurate graphs of these functions or to sketch their shapes, showing important values on the axes.

The shape of the graphs can also help you to find the values of x that satisfy equations such as $\sin x° = 0.5$.

On the calculator, finding the inverse sine of 0·5 will

> ### Exam tip
> Learn the shapes of the basic graphs of
> $$y = \cos x°$$
> $$y = \sin x°$$
> $$y = \tan x°.$$

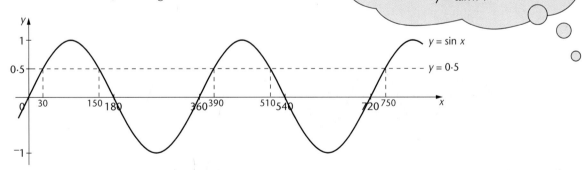

give you the answer of 30°. However, the graph shows that there is an infinite number of solutions. Use the symmetry of the graph to see that ⁻210°, 150°, 390°, 510°, etc. are other solutions. You can use your calculator as a check to see that the sine of all these angles is 0·5.

Chapter 14 *Trigonometrical functions*

EXAMPLE 1

Sketch the graph of $y = \cos x$ for values of x from 0 to 360°. Use the graph and your calculator to find two values of x between 0 and 360° for which $\cos x = {}^-0.8$. Give your answers to 1 d.p.

From the calculator, $\text{invcos}({}^-0.8) = 143.1°$

$180 - 143.1 = 36.9$

From the symmetry of the graph, the other solution is $x = 180 + 36.9 = 216.9°$

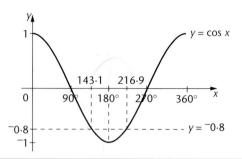

EXERCISE 14.1A

1 Draw accurately the graph of $y = \sin x$ for values of x from 0° to 360°, plotting values every 10°. Use a scale of 1 cm to 20° for x and 2 cm to 1 unit for y. Use your graph to solve for $0 \le x \le 360°$.
   a) $\sin x = 0$ **b)** $\sin x = 0.45$ **c)** $\sin x = {}^-0.60$

2 Sketch the graph of $y = \cos x°$ for values of x from $^-180°$ to 360°. Use the graph and your calculator to find, to 1 d.p., the three solutions of $\cos x = 0.3$ between $^-180°$ and 360°.

3 With the help of a sketch graph, or otherwise, find two solutions of $\tan x = {}^-2$ between 0° and 360°. Give your answers to the nearest degree.

4 On the same diagram, sketch the graph of $y = \sin x°$ and $y = \cos x°$ for values of x between 0° and 360°. State the values of x between 0° and 360° for which $\sin x° = \cos x°$.

5 One solution of $\cos x = 0.5$ is $x = 60°$. Without using a calculator, use the symmetry of the graph $y = \cos x$ to find the four solutions of $\cos x = {}^-0.5$ between 0 and 720°.

6 Give another two angles which have the same sin value as
   a) $\sin 20°$ **b)** $\sin 120°$ **c)** $\sin {}^-45°$ **d)** $\sin 390°$.

7 Give three other angles which have a sin value equal to
   a) $\sin 40°$ **b)** $\sin {}^-80°$.

8 Give three other angles which have a cos value equal to
   a) $\cos 140°$ **b)** $\cos {}^-120°$.

9 Give three other angles which have a tan value equal to
   a) $\tan 45°$ **b)** $\tan 120°$.

10 For $0° \le x \le 360°$ solve:
   a) $\sin x = {}^-0.37$ **b)** $\cos x + 2 = 3$ **c)** $\sin x = \tan x$ **d)** $2 \sin x = 1$.

EXERCISE 14.1B

1 Draw accurately the graph of
$y = \tan x$ for values of x from $0°$ to
$360°$, plotting values every $10°$. Use a
scale of $1\,\text{cm}$ to $20°$ for x and $2\,\text{cm}$ to
1 unit for y, scaling y from $^-4$ to 4.

Use your graph to solve, for
$0° < x < 360°$:

a) $\tan x = 1$ **b)** $\tan x = 2$.

2 Sketch the graph of $y = \sin x°$ for
values of x from 0 to $540°$. Use your
graph and calculator to find four
solutions of $\sin x = 0\!\cdot\!8$. Give your
answers to 1 d.p.

3 With the help of a sketch graph,
or otherwise, find two solutions of
$\sin x = ^-0\!\cdot\!2$ between 0 and $360°$. Give
your answers to the nearest degree.

4 One value of x for which $\cos x° = ^-0\!\cdot\!3$
is 107, to the nearest integer.
Without using your calculator, use
the symmetry of the graph $y = \cos x°$
to find two other solutions of this
equation between 0 and 540.

5 One solution of $\tan x = 1$ is $x = 45°$.
Without using a calculator, use the
symmetry of the graph $y = \tan x$ to
find the two solutions of $\tan x = ^-1$
between $0°$ and $360°$.

6 Give another two angles which have
the same cos value as

a) $\cos 40°$ **b)** $\cos 90°$ **c)** $\cos 285°$.

7 Give three other angles which have a
tan value equal to

a) $\tan 40°$ **b)** $\tan\ ^-80°$.

8 Give three other angles which have a
sin value equal to

a) $\sin 130°$ **b)** $\sin\ ^-120°$.

9 Give three other angles which have a
tan value equal to

a) $\tan 30°$ **b)** $\tan 135°$.

10 For $0° \leq x \leq 360°$ solve:

a) $\tan x = ^-1\!\cdot\!56$ **b)** $\sin x + 2 = 1\!\cdot\!5$

c) $\cos x = \sin x$ **d)** $3 \cos x = 2$.

ACTIVITY 2

Use a graphics calculator or a graph-drawing computer
program for this activity.

1 Draw the graph of $y = \sin x$ for values of x from
$^-180°$ to $540°$.

On the same axes, rescaling as necessary, draw the
graphs of: $y = 2 \sin x$

$y = 3 \sin x$

$y = ^-2 \sin x$

$y = 0\!\cdot\!5 \sin x$

Experiment further till you think you know what
the graph of $y = a \sin x$ looks like for any value of
a. Can you say how the graph of $y = \sin x$ has
been transformed?

2 Clear the screen and try the graphs of $y = \cos x$,
$y = 2 \cos x$, etc. Can you describe what happens?
What about $y = \tan x$, $y = 2 \tan x$, etc.?

3 Clear the screen and draw again the graph of
$y = \sin x$ for values of x from $^-180°$ to $540°$.

Chapter 14 *Trigonometrical functions*

Activity 2 cont'd

On the same axes, draw the graph of $y = \sin 2x$. This notation means $y = \sin(2x)$, but you will probably not need the brackets for your program. Draw also the graphs of $y = \sin 3x$, $y = \sin 0.5x$, $y = \sin(^-x)$

Experiment further till you think you know what the graph of $y = \sin ax$ looks like for any value of a.

4 What about the graphs of $y = \cos ax$ or $\tan ax$? Can you describe how the graphs of $y = \cos x$ or $\tan x$ are transformed to these graphs?

Other trigonometrical graphs

If, instead of plotting $y = \sin x°$, you plot $y = 3 \sin x°$, what difference does it make?

Here are both of these graphs plotted together.

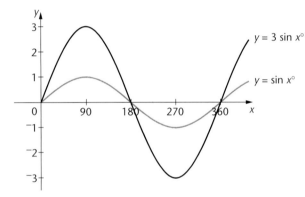

The period has stayed the same, 360, but the amplitude has increased from 1 to 3.

Similarly the graph of $y = 5 \cos x°$ has a period of 360 but an amplitude of 5.

The graph of $y = \cos 3x°$ is different. $\cos 3x$ means $\cos(3x)$.

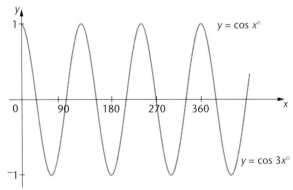

Here are the graphs of $y = \cos x°$ and $y = \cos 3x°$ plotted together for comparison.

$y = \cos 3x$ has an amplitude of 1 but a period of $\dfrac{360}{3} = 120°$.

The graphs of $y = a \sin bx$ and $y = a \cos bx$ each have amplitude a and period $\dfrac{360}{b}$.

Chapter 14 *Trigonometrical functions*

EXAMPLE 2

Sketch the graph of $y = 3 \sin 2x°$ for $x = 0$ to 360. Find the solutions of $3 \sin 2x° = 2$ between 0 and 90, giving your answer to 1 d.p.

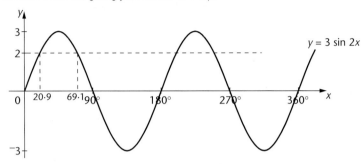

$$3 \sin 2x° = 2$$
$$\sin 2x° = \frac{2}{3}$$

From the inverse sine function on the calculator, $2x = 41·810 \ldots$

$$x = 20·9 \text{ to } 1 \text{ d.p.}$$

From the symmetry of the graph, the other solution is $90 - 20·9 = 69·1$.

EXERCISE 14.2A

1   Draw accurately the graph of $y = 2 \sin x°$ for values of x from 0° to 180°, plotting values every 10°.

2   Draw accurately the graph of $y = \cos 2x$ for values of x from 0° to 180°, plotting values every 10°.

3   State the amplitude of

   a) $y = 3 \sin x$ **b)** $y = 4 \cos 2x$ **c)** $y = 2 \cos 0·5x$.

4   State the period of each of the graphs in question 3.

5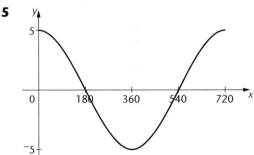

This is part of the graph of $y = a \cos bx$.
State the values of a and b.

Exercise 14.2A cont'd

6

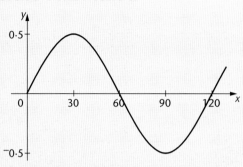

This is part of the graph of
$y = a \sin bx$. State the values of a and b.

7 Sketch the graph of $y = \sin \frac{1}{2}x$ for values of x from $0°$ to $360°$.

8 Find the four solutions of $\sin 2x = 0.5$ between $0°$ and $360°$.

9 Find the solutions of $\cos 2x° = 0$ for $0 < x \leq 360$.

10 On the same diagram, sketch the curves of $y = \cos 2x$ and $y = \sin x$ for $0° \leq x \leq 90°$. How many solutions of the equation $\cos 2x = \sin x$ are there for $0° \leq x \leq 90°$?

EXERCISE 14.2B

1 Draw accurately the graph of $y = 3 \cos x°$ for values of x from $0°$ to $180°$, plotting values every $10°$.

2 Draw accurately the graph of $y = \sin 4x$ for values of x from $0°$ to $180°$, plotting values every $10°$.

3 State the amplitude of
  a) $y = 5 \cos x$ **b)** $y = 2 \sin 3x$
  c) $y = 4 \sin \frac{1}{3}x$.

4 State the period of each of the graphs in question 3.

5

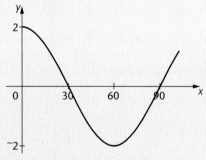

This is part of the graph of
$y = a \cos bx$. State the values of a and b.

6 This is part of the graph of
$y = a \sin bx$. State the values of a and b.

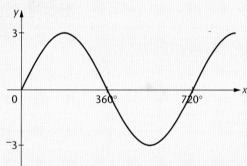

7 Sketch the graph of $y = \cos \frac{1}{3}x$ for values of x from $0°$ to $540°$.

8 Find all the solutions of $\cos 3x = {}^{-}1$ between $0°$ and $360°$.

9 Find the solutions of $2 \sin x° = 1$ for $0 < x \leq 360$.

10 On the same diagram, sketch the curves $y = \sin 3x$ and $y = \cos x$ for $0° \leq x \leq 90°$. How many solutions of the equation $\sin 3x = \cos x$ are there for $0° \leq x \leq 90°$?

Key ideas

● The shapes and main features of trigonometrical graphs:

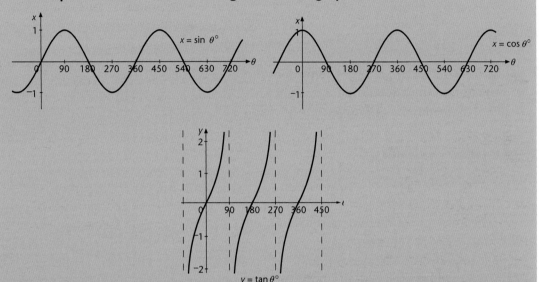

● Your calculator and the symmetry of these graphs will help you to find solutions to trigonometrical equations.

● The 'length' of a repeating pattern is called the **period**. For a wave, the amount it varies from its mean is called the **amplitude**.

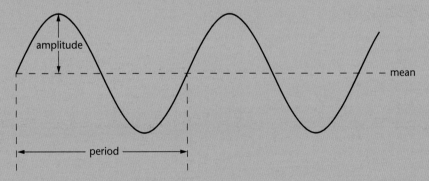

● The graphs of $y = a \sin bx$ and $y = a \cos bx$ each have amplitude a and period $\frac{360}{b}$.

Chapter 14 *Trigonometrical functions*

15 Transforming functions

You should already know

- the basic transformations of reflections and translations
- how to find the equation of a straight line
- the shapes of basic graphs such as $y = x^2$, $y = x^3$, $y = \sin x$.

Function notation

$y = f(x)$ means that y is a function of x.

If $y = 4x - 3$ then $f(x) = 4x - 3$.

$f(2)$ means the value of the function when $x = 2$.

In this case, $f(2) = 4 \times 2 - 3 = 5$.

Function notation is a useful shorthand when several different functions are being described. As well as $f(x)$, $g(x)$ and $h(x)$ are commonly used to describe functions.

EXAMPLE 1

If $f(x) = 3x^2 - 5$, find **a)** $f(2)$ **b)** $f(^-1)$

a) $f(2) = 3 \times 2^2 - 5 = 7$

b) $f(^-1) = 3 \times (^-1)^2 - 5 = {}^-2$

EXAMPLE 2

If $g(x) = 5x + 6$, **a)** solve $g(x) = 8$ **b)** write an expression for **(i)** $g(3x)$ **(ii)** $3g(x)$.

a) $g(x) = 8$

$\quad 5x - 6 = 8$

$\quad 5x \quad = 2$

$\quad x \quad = 0.4$

b) (i) $g(3x) = 5(3x) + 6 = 15x + 6$

$\quad$ **(ii)** $3g(x) = 3(5x + 6) = 15x + 18$

ACTIVITY

As you go, sketch (or print out) the graphs and make a note of your results. (Hint: make sure each curve on your sketch or print out is labelled clearly with its equation.) Use a graph-drawing program to plot the graphs. When the grid gets too crowded, start on a new set of axes. At least, start each section with a fresh set of axes.

Consider first the function $f(x) = x^2$.

$y = f(x) + a$

1   Plot the graph of $y = f(x)$.

2   On the same axes, plot the graph $y = x^2 + 2$. This can be called the graph of $y = f(x) + 2$.

3   What transformation do you have to do to the first graph to obtain the second?

4   Try some other graphs in the form $y = x^2 + a$, where a can take any value (try positive, negative, integers, fractions ...).

5   Describe the transformation from the graph of $y = f(x)$ to $y = f(x) + a$, for any a.

$y = f(x + a)$.

6   Try plotting $y = f(x + 1)$ (i.e. the graph of $y = (x + 1)^2$).

7   What about $f(x - 2)$ or $f(x + 2)$?

8   Experiment until you can describe the transformation of $y = f(x)$ to obtain $y = f(x + a)$.

$y = af(x)$.

9   Plot the graph of $y = 2f(x)$ (i.e. $y = 2x^2$).

10 Try other graphs of the form $y = ax^2$. Experiment until you can describe the transformation that maps $y = f(x)$ onto $y = af(x)$.

$y = f(ax)$

11 Plot $y = f(2x)$ (i.e. $y = (2x)^2$).

12 Try $y = f\left(\frac{x}{2}\right)$ and $f(^-3x)$ etc.

13 Describe the transformation that maps $y = f(x)$ onto $y = f(ax)$.

What happens if you change the function?

Work through each section again, using $y = \sin x$ for $^-360 \leq x \leq 360°$ instead of $y = x^2$.

What differences are there?

What similarities are there?

Have you found any 'special cases'?

What about other functions ...?

Could you combine some of these transformations and state the equation of the resulting graph?

Other ideas?

Translations

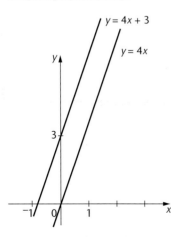

You already know that the graph of $y = 4x$ is a straight line through the origin with gradient 4. The graph of $y = 4x + 3$ is a straight line with gradient 4 through the point (0, 3). You can think of this as a transformation by saying that the first graph has been translated by $\binom{0}{3}$ to get the second.

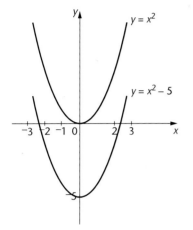

Similarly, the graph of $y = x^2 - 5$ is the same shape as the graph of $y = x^2$, translated $\begin{pmatrix} 0 \\ -5 \end{pmatrix}$.

The same thing is true for all families of graphs.

> The graph of $y = f(x) + a$ is the graph of
> $y = f(x)$ translated by $\begin{pmatrix} 0 \\ a \end{pmatrix}$.

Translating parallel to the x axis, a different pattern emerges.

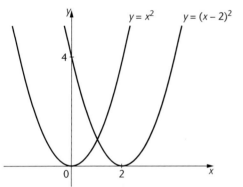

The equation $(x - 2)^2 = 0$ has root $x = 2$. Looking at the graphs of $y = x^2$ and $y = (x - 2)^2$, it is clear that $y = x^2$ has been translated $\begin{pmatrix} 2 \\ 0 \end{pmatrix}$.

> In general:
> the graph of $y = f(x - a)$ is the graph of
> $y = f(x)$ translated by $\begin{pmatrix} a \\ 0 \end{pmatrix}$.

Exam tip

It is not necessary to learn all these and the later results. You must understand them so that you can work them out when needed.

EXAMPLE 3

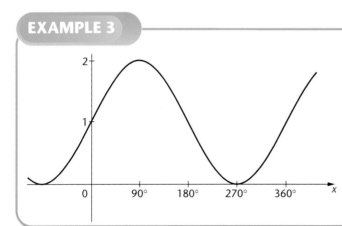

State the equation of the sine curve drawn here.

The curve of $y = \sin x$ has been translated by $\begin{pmatrix} 0 \\ 1 \end{pmatrix}$.

So its equation is $y = \sin x + 1$.

EXERCISE 15.1A

1 Sketch on the same diagram the graphs of

a) $y = x^2$ **b)** $y = x^2 + 3$.

State the transformation which maps **a)** onto **b)**.

2 Sketch on the same diagram the graphs of

a) $y = x^2$ **b)** $y = (x + 2)^2$.

What transformation maps **a)** onto **b)**?

3 Sketch on the same diagram the graphs of

a) $y = x^2$ **b)** $y = (x - 2)^2$

c) $y = (x - 2)^2 + 3$.

What transformation maps **a)** onto **c)**?

4 Sketch the result of translating the graph of $y = \sin x$ by $\begin{pmatrix} 0 \\ -1 \end{pmatrix}$. State the equation of the transformed graph.

5 State the equation of the graph of $y = x^2$ after it has been translated by

a) $\begin{pmatrix} 0 \\ -5 \end{pmatrix}$ **b)** $\begin{pmatrix} -2 \\ 0 \end{pmatrix}$.

6

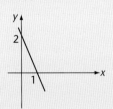

This is the graph of $y = f(x)$. Sketch the graphs of

a) $y = f(x) - 2$

b) $y = f(x - 2)$.

7

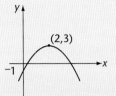

This is the graph of $y = g(x)$. State the coordinates of the maximum point on the graphs of

a) $y = g(x + 1)$

b) $y = g(x) + 1$.

8 State the equation of the graph of $y = x^2$ after it has been translated by $\begin{pmatrix} 1 \\ 2 \end{pmatrix}$.

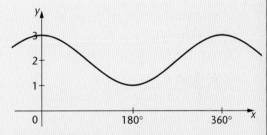

9 This graph is a transformed cosine curve.

State its equation.

10 The graph of $y = x^2$ is translated by $\begin{pmatrix} -2 \\ 3 \end{pmatrix}$.

a) State the equation of the transformed graph.

b) Show that this equation may be written as $y = x^2 + 4x + 7$.

EXERCISE 15.1B

1 Sketch on the same diagram the graphs of

a) $y = {}^-x^2$ **b)** $y = 2 - x^2$.

State the transformation which maps **a)** onto **b)**.

2 Sketch on the same diagram the graphs of

a) $y = {}^-x^2$ **b)** $y = -(x + 3)^2$.

What transformation maps **a)** onto **b)**?

3 Sketch on the same diagram the graphs of

a) $y = x^2$

b) $y = (x + 2)^2$

c) $y = (x + 2)^2 - 3$.

What transformation maps **a)** onto **c)**?

4 Sketch the result of translating the graph of $y = \cos x$ by $\binom{0}{1}$. State the equation of the transformed graph.

5 State the equation of the graph of $y = \tan x$ after it has been translated by

a) $\binom{3}{0}$ **b)** $\binom{0}{4}$.

6

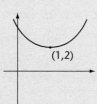

This is the graph of $y = f(x)$. State the coordinates of the minimum point on the graph of

a) $y = f(x) - 2$ **b)** $y = f(x + 2)$.

7

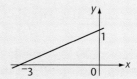

This is the graph of $y = g(x)$. Sketch the graphs of

a) $y = g(x) - 3$

b) $y = g(x + 3)$.

8 State the equation of the graph of $y = x^2$ after it has been translated by $\binom{3}{-4}$.

9 This graph is a transformed sine curve.

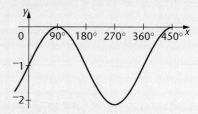

State its equation.

10 Show that the equation $y = x^2 - 6x + 1$ may be written as $y = (x - 3)^2 - 8$.

Hence state the coordinates of the minimum point on the graph of $y = x^2 - 6x + 1$.

One-way stretches

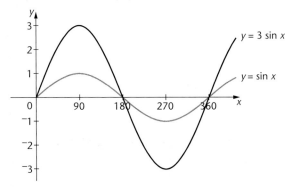

In Chapter 14 you learned about the shapes of sine and cosine graphs.

This diagram shows the graphs of $y = \sin x°$ and $y = 3 \sin x°$.

To get from $y = \sin x$ to $y = 3 \sin x$, the graph has been stretched parallel to the y axis with scale factor 3. This is an example of the general principle:

> The graph of $y = kf(x)$ is a one-way stretch of the graph of $y = f(x)$ parallel to the y axis with scale factor k.

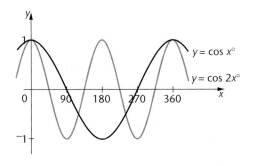

The next graph shows the graphs of $y = \cos x°$ and $y = \cos 2x°$.

For the graph of $y = \cos 2x$ compared with the graph of $y = \cos x$, twice as much curve has been squashed into each part of the x axis. This is described formally as a one-way stretch parallel to the x axis with scale factor $\frac{1}{2}$. This is an example of the general principle:

> The graph of $y = f(kx)$ is a one-way stretch of the graph of $y = f(x)$ parallel to the x axis with scale factor $\frac{1}{k}$.

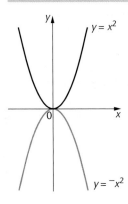

When $k = {}^-1$ in these one-way stretches, there is a much simpler way of describing the transformation – as reflections. For example, the graph of $y = {}^-x^2$ is a reflection of $y = x^2$ in the x axis.

> The graph of $y = {}^-f(x)$ is a reflection in the x axis of the graph of $y = f(x)$.

Comparing the graphs of $y = x^3 + 1$ and $y = {}^-x^3 + 1$ [or $y = ({}^-x)^3 + 1$], you can see that one is a reflection of the other in the y axis.

> The graph of $y = f({}^-x)$ is a reflection in the y axis of the graph of $y = f(x)$.

Exam tip

Take care not to confuse $f({}^-x)$ with ${}^-f(x)$.

255

EXAMPLE 4

If $f(x) = x^2 + 2$, state in the form of $y = ax^2 + bx + c$ the equation of **a)** $y = f(4x) + 2$ **b)** $y = 3f(x)$.

a) $y = (4x)^2 + 2$ $\quad y = 16x^2 + 2$

b) $y = 3(x^2 + 2)$ $\quad y = 3x^2 + 6$

EXERCISE 15.2A

1 Sketch on the same diagram the graphs of $y = \cos x$ and $y = 2\cos x$ for $0 \le x \le 360°$.

Describe the transformation that maps $y = \cos x$ onto $y = 2\cos x$.

2 Sketch on the same diagram the graphs of $y = \sin x$ and $y = \sin \frac{1}{2}x$ for $0° \le x \le 360°$.

Describe the transformation that maps $y = \sin x$ onto $y = \sin \frac{1}{2}x$.

3 Describe the transformation that maps $y = \sin x$ onto $y = \sin 3x$.

4 Describe the transformation that would map:

a) $y = \sin x + 1$ onto $y = \sin(^-x) + 1$

b) $y = x^2 + 2$ onto $y = ^-x^2 - 2$

c) $y = x^2$ onto $y = 3x^2$.

5 The graph of $y = \cos x$ is transformed by a one-way stretch parallel to the x axis with scale factor $\frac{1}{3}$. State the equation of the resulting graph.

6 State the equation of the graph of $y = x^2 + 5$ after:

a) reflection in the y axis,

b) reflection in the x axis.

7 State the equation of the graph of $y = x + 2$ after:

a) a one-way stretch parallel to the y axis with scale factor 3

b) a one-way stretch parallel to the x axis with scale factor $\frac{1}{2}$.

8 Describe transformations to map $y = g(x)$ onto:

a) $y = g(x) + 1$ **b)** $y = 3g(x)$
c) $y = g(2x)$ **d)** $y = 5g(3x)$.

9 State the equation of the graph of $y = 2x + 1$ after:

a) reflection in the x axis

b) reflection in the y axis

c) a one-way stretch parallel to the x axis with scale factor 0·5.

10 The graph of $y = x^2$ is stretched parallel to the x axis with scale factor 2.

a) State the equation of the resulting graph.

b) What point does (1, 1) map onto under this transformation?

c) (i) What is the scale factor of the stretch parallel to the y axis which maps $y = x^2$ onto the same graph?

(ii) What point does (1, 1) map onto under this transformation?

Exercise 15.2A cont'd

11

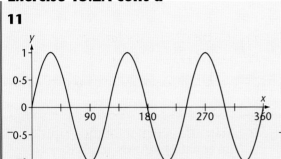

The equation of this graph is
$y = \sin ax$. Find a.

12

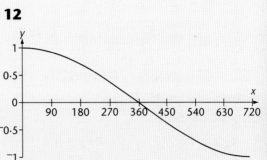

The equation of this graph is
$y = \cos bx$. Find b.

EXERCISE 15.2B

1 Sketch on the same diagram the graphs of $y = \cos x$ and $y = {}^-\cos x$ for $0 \le x \le 360°$.
Describe the transformation that maps $y = \cos x$ onto $y = {}^-\cos x$.

2 Sketch on the same diagram the graphs of $y = \sin x$ and $y = \sin 2x$ for $0 \le x \le 360°$.
Describe the transformation that maps $y = \sin x$ onto $y = \sin 2x$.

3 Describe the transformation that maps $y = \sin x$ onto $y = \sin \frac{1}{3}x$.

4 Describe the transformation that would map:

a) $y = \cos x + 1$ onto $y = {}^-\cos x - 1$ **b)** $y = x + 2$ onto $y = {}^-x + 2$

c) $y = x^2$ onto $y = 5x^2$.

5 The graph of $y = \sin x$ is transformed by a one-way stretch parallel to the x axis
with scale factor $\frac{1}{4}$.
State the equation of the resulting graph.

6 State the equation of the graph of $y = x^2 - 1$ after:

a) reflection in the y axis,   **b)** reflection in the x axis.

Exercise 15.2B cont'd

7 State the equation of the graph of $y = 4x + 1$ after:

a) a one-way stretch parallel to the y axis with scale factor 4

b) a one-way stretch parallel to the x axis with scale factor 0·5.

8 Describe transformations to map $y = h(x)$ onto:

a) $y = h(x) - 2$ **b)** $y = 3h(x)$
c) $y = h(0·5x)$ **d)** $y = 4h(2x)$.

9 State the equation of the graph of $y = x^2 + 3$ after:

a) reflection in the x axis

b) reflection in the y axis

c) a one-way stretch parallel to the x axis with scale factor 0·5.

10 Use the shape of the graph of $y = \sin x$ to sketch on the same axes the graphs for $0° \leq x \leq 180°$ of

a) $y = \sin \frac{1}{2}x$

b) $y = 3 \sin \frac{1}{2}x$.

11

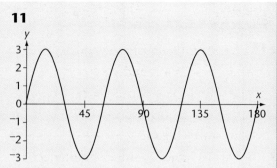

The equation of this graph is $y = a \sin bx$. Find a and b.

12

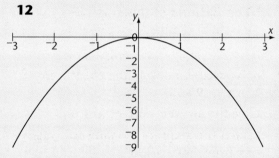

The equation of this graph is $y = ax^2$. Find a.

Key ideas

- The graph of $y = f(x) + a$ is the graph of $y = f(x)$ translated by $\begin{pmatrix} 0 \\ a \end{pmatrix}$.

- The graph of $y = f(x - a)$ is the graph of $y = f(x)$ translated by $\begin{pmatrix} a \\ 0 \end{pmatrix}$.

- The graph of $y = kf(x)$ is a one-way stretch of the graph of $y = f(x)$ parallel to the y axis with scale factor k.

- The graph of $y = f(kx)$ is a one-way stretch of the graph of $y = f(x)$ parallel to the x axis with scale factor $\frac{1}{k}$.

- The graph of $y = {}^-f(x)$ is a reflection in the x axis of the graph of $y = f(x)$.

- The graph of $y = f({}^-x)$ is a reflection in the y axis of the graph of $y = f(x)$.

D1 Revision exercise

1 Solve simultaneously the equations:

a) $y = x^2 - 2x + 3$ and $y = 2x$

b) $y = 2x^2 - 3x + 3$ and $y = 3x - 1$

c) $y = x^2 - 4x + 5$ and $y + 4x = 6$

d) $x^2 + y^2 = 36$ and $y = x + 6$.

2 Solve algebraically these simultaneous equations.

a) $y = x^2 - 3x$ and $y = 8 - x$

b) $y = 2x^2 - 4x + 1$ and $y = 3x - 2$

c) $y = x^2 - 5x + 5$ and $x - 2y = 5$

d) $x^2 + y^2 = 29$ and $x + 2y = 1$

3 Solve simultaneously the equations $x^2 + y^2 = 9$ and $y = x + 2$.

Give the answers correct to 1 d.p.

4 Sketch the graph of $y = \tan x$ for values of x from $^-90°$ to $450°$.

5 For what angles between $0°$ and $360°$ does $\tan x = 1$?

6 Sketch the graph of $y = \cos x$ for $0° \leq x \leq 360°$. Given that one solution of $\cos x = {}^-0.8$ is $143°$ to the nearest degree, find the other solution between $0°$ and $360°$.

7 Given that one solution of $\sin x = \dfrac{-1}{2}$ is $x = {}^-30°$, use the symmetry of the graph of $y = \sin x$ to find the solutions between $0°$ and $360°$.

8 Using a calculator and sketch graph, or otherwise, solve the equation $\cos x = 0.2$ for $0 \leq x \leq 360°$.

9 On the same set of axes, sketch the graphs of $y = \cos x°$ and $y = \cos 2x°$ for $0 \leq x \leq 360$.

10 For $0 \leq x \leq 360$, for what values of x does $\sin 2x° = 1$?

11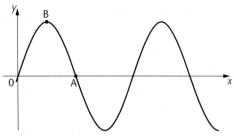

This is the graph of $y = 2 \sin 3x°$. State the coordinates of A and B.

12 Sketch the graph of $y = 3 \cos 2x$ for $0° \leq x \leq 360°$.

13 $f(x) = x^2 - 2$. Find the equation of the graph of $y = x^2 - 2$ after it has been translated by $\begin{pmatrix} 1 \\ 0 \end{pmatrix}$.

14 Describe the transformation which would map $y = g(x)$ onto the graph of

a) $y = g(3x)$ **b)** $y = 4g(x)$ **c)** $y = g(^-x)$.

15 Sketch the graph of $y = \sin(4x)°$ for values of x from 0 to 100.

16 a) Sketch the graph of $y = \sin(x + 90)°$.

b) State the equation of this graph more simply.

17 State the equation of the graph $y = \cos x$ after

a) a translation of $\begin{pmatrix} 0 \\ 3 \end{pmatrix}$

b) a one-way stretch parallel to the x axis with scale factor 0·25.

18 The graph of $y = x^2$ is translated by $\begin{pmatrix} 3 \\ 1 \end{pmatrix}$ and then stretched with scale factor 2 parallel to the y axis.

a) Find the equation of the resulting curve.

b) Find the coordinates of the minimum point on this curve.